Praise for *Chilled*

… A chill-cabinet of curiosities: hot stuff, and deeply cool …

The Spectator

I can't think of a better light non-fiction summer read than this.

Independent

Buoyant, idiosyncratic and very funny … this history of what is, ultimately, a rather mundane piece of kitchenware is consistently fascinating. Cool story.

Financial Times

Fun and eye-opening … this is an inspiring, compelling and utterly convincing book.

The Sunday Times

Jackson sees the appliance as 'humanity's greatest achievement' … *Chilled* attests to his abilities as a historian and a bit of a comedian.

Times Literary Supplement

… Plenty of fascinating stuff.

The Times

Jackson handles tricky ideas deftly … like a well-stocked refrigerator, this book is packed with tasty morsels.

BBC Focus

In his entertaining new book, *Chilled*, Jackson walks us through the creation of cold – or, at least, man-made cold. He explains how frigid air made all sorts of things possible, from the variety of food we eat to the hydrogen bomb.

The Washington Post

… A nutritious little book.

The Daily Mail

Without refrigeration, this delightfully illuminating book reminds us, not only would there be no ice cream or cold lager, there would be no MRI scanners in hospitals, no super-computers, no weekly food shop.

Mail on Sunday

A NOTE ON THE AUTHOR

Tom Jackson is a science writer based in Bristol, UK. Tom specialises in recasting science and technology into lively historical narratives. After almost 20 years of writing, Tom has uncovered a wealth of stories that help to bring technical content alive and create new ways of enjoying learning about science.

In his time, Tom has been a zoo keeper, travel writer, buffalo catcher and filing clerk, but he now writes for adults and children, for books, magazines and TV.

CHILLED

How refrigeration changed the world, and might do so again

Tom Jackson

Bloomsbury Sigma
An imprint of Bloomsbury Publishing Plc

50 Bedford Square
London
WC1B 3DP
UK

1385 Broadway
New York
NY 10018
USA

www.bloomsbury.com

BLOOMSBURY and the Diana logo are trademarks of Bloomsbury Publishing Plc

First published 2015. Paperback edition 2016.

British Library Cataloguing-in-Publication Data
A catalogue record for this book is available from the British Library.

Library of Congress Cataloguing-in-Publication data has been applied for.

ISBN (hardback) 978-1-4729-1143-8
ISBN (paperback) 978-1-4729-1144-5
ISBN (ebook) 978-1-4729-1142-1

2 4 6 8 10 9 7 5 3 1

Typeset in 12pt Bembo Std by Deanta Global Publishing Services, Chennai, India

Printed and bound in Great Britain by CPI Group (UK) Ltd, Croydon CR0 4YY

Bloomsbury Sigma, Book Five

For Sarah

Contents

Introduction

The refrigerator is something of a Boo Radley, the cloistered figure in Harper Lee's *To Kill a Mockingbird*: it's normally pale, frequently indoors, seldom thought about much but always there, and in the end (spoiler alert) we need it to make everything all right.

The point is that among the cast of characters that tell the story of refrigeration and its impact on humanity, the refrigerator itself does not have much of a starring role. It just hangs around in the background most of the time, being cool.

You'd be forgiven for thinking that refrigerators, fridges, are nothing to get excited about. They are nothing special. There are hundreds of millions of them. Every household in the developed world has one; a quarter of American homes have at least two. And elsewhere a refrigerator is at the top of the wish list of household appliances – just below the television. When we fancy something cold – an ice cream, glass of milk, chilled beer – there they are in the refrigerator. In this day and age, access to cold is so easy there's no need to think about it.

Nicolas Monardes, a Spaniard writing in 1574, reports more of a difficulty. Cold was available to him for the right price as ice, but Monardes advises that care should be taken to avoid sources mingled with 'smelles and evil smokes … especially that which passeth by places where are rotten plantes, and naughtie tree, and where dead babies are.' All good advice then and good advice now, but thankfully refrigeration has moved on.

The fridge occupies a central place in the lives of the privileged many. To the initiated, life without one would be intolerable. By following just a few simple rules (helpfully spelled out on packets) we can enjoy foods and drinks of pristine freshness whenever we desire them. And when we run out, we simply make a trip to the larger refrigerators at the supermarket to restock.

So ubiquitous is this routine that we don't give it a second thought. We do not consider that our kitchen fridge is the very tip of a chilled tendril, one of millions more that make up a network known in food-industry circles as the cold chain. The cold chain, with its myriad nodes and branches, entangles the globe, creating a temperature-controlled transport corridor that connects the farmer's field and the trawler's hold to every grocery store chiller.

Linking to the chain gets us sashimi in Vegas, strawberries at Christmas and sorbets whatever the weather. The chain gives us choice and the luxury of time to make it. No longer is fresh produce rushed into cities during the cool of the night to be consumed within hours of its arrival. Forget the skyscrapers, subways and information superhighways – it's the fridge that makes a modern city. The refrigerators of Greater Tokyo, the world's largest urban area, provide the ingredients for at least 113 million meals a day. Without the cold chain, life in such a metropolis would be an unimaginable ordeal.

But the modern world is full of innovations, so what's the big deal about the refrigerator? The kitchen alone is equipped with all kinds of technological necessities – the electric kettle, the fan oven, the microwave … and the fridge. There is of course one obvious defining difference: all but one of them heats, and only the fridge cools.

Heat and light have been at the command of humanity for at least 100 millennia since we learned to make and control fire. We only really won the battle over cold a century ago, and still the spoils of that victory have yet to be shared with many parts of the globe. Today, the idea that hot and cold are two sides of the same coin seems second nature, but that was far from obvious to the long line of scholars whose combined efforts figured it out for us. Their stories run from the inspired to the bizarre, as they employed gnomes, meteors, perpetual motion machines and tortured mice to reveal the truth.

The knowledge that they – the likes of Cornelius Drebbel, Robert Boyle and James Joule – revealed was to form the basis of thermodynamics, the field of physics that tackles the behaviour of energy. Martin Goldstein, an American chemist and author on the subject of thermodynamics, had it right when he wrote: 'Some people want to know how a refrigerator works. Others want to know the fate of the universe ... The science that relates them is thermodynamics.'

A refrigerator is a 'heat pump', which on the face of it is perhaps an uninspiring term. However, dig a little deeper into the concept and it reveals something rather amazing – tiny acts of rebellion against the conformity of the universe.

The idea of a 'heat sink' is perhaps a familiar one. Basically it means that the energy of a hot space is transferred to one that is less hot. So heat pours out of the Sun, warming objects around it – mostly the planets, including Earth. In turn Earth's heat energy dissipates into the growing emptiness of space, which is the ultimate 'heat sink'. This one-way travel is an unbreakable law of thermodynamics, but one that can be disrupted, albeit temporarily, by a 'heat pump'. A heat pump can push heat against the universal flow. In the case of a refrigerator, it is pushing heat out of the food and freezer compartments into the surroundings, and as a result everything inside gets colder.

Our planet alone, among the billions postulated to exist, is home to an army of heat pumps, machines that are working against the inexorable collapse of order into chaos. Now it is perhaps becoming clearer why palaeolithic man had little trouble torching a wooden stick, but had to wait several dozen millennia before he could put an ice-lolly on it.

The technology that could harness physical laws to create cold was far from an overnight sensation. It took 170 years to transform it from a prototype in 1750 to anything approaching a mass-market product. However, the demand for cold had raced ahead of the technology.

The storage of natural ice had been a niche practice for the wealthiest since antiquity, especially in Asia. As with so many other technologies, Eastern know-how was put to work in Renaissance Europe, so the nobility of Italy, France and Spain also had the luxury of chilled wines and iced desserts. Luxury was not all it was used for. In 1503, Cesare Borgia, the notorious leader of an army that had swept across swathes of southern Europe in the name of the Pope – Cesare was a bastard son of the incumbent – used his supply of ice for a less traditional purpose.

Cesare had already made his mark on history by that time. He is reputed to be the inspiration behind Niccolò Machiavelli's *The Prince* (1513), a handy guide on how to rule with terror and treachery, and he also appears as one of the bad guys in the Assassin's Creed games. Neither makes reference to his possibly unique experience with ice. Both he and his father, Pope Alexander VI, were struck down by a terrible fever – retold as a planned poisoning gone awry. While the Pope opted for a thorough-going bleeding to combat the malady, Cesare's fever raged so strongly that he was immersed in a man-sized oil jar filled with iced water. It did the trick; Cesare survived to lie, cheat and murder for another 20 years – but the treatment is reported to have made all of his skin fall off (Pope Alexander died, and his body was so swollen by the disease that it had to be literally hammered into the coffin).

The medicinal virtues of cold were widely considered – although drinking too much cold water was also deemed a dangerous and unnatural act – but preservation of food was not regarded as a primary use for ice at first. The rise of icehouses across Europe was primarily due to a demand for chilled wines and dessert but the vogue grew for stocking them with sides of fresh meat and ripening fruits.

It would take some nineteenth-century American get-up-and-go to commercialise the use of ice on a global scale, first as slabs of natural ice and then through mechanical refrigeration. A social revolution had begun, slow for sure but unstoppable in its impact on food and society as a whole. Kelvinator, a leading brand in the early days, put it this way in *For the Hostess*, their 1920s recipe book for the first refrigeration generation: 'The housewife sees her labors lightened, sees more hours of leisure, and with it all, extraordinary economies.'

What's changed? Nothing really, although we wouldn't use the term 'housewife' any more, and that is surely no coincidence.

The refrigerator has changed the way we live. That was already obvious in 1931, judging by a magazine article titled 'The New Ice Age':

> *If the stupendous system of food preservation and transportation which supports us were interfered with, even for short time, our present daily existence would become unworkable, Cities with thousands of inhabitants would fade away. We would probably turn into beasts in our frantic struggles to reach the source of supply … It is not extravagant to say that our present form of civilisation is dependent upon refrigeration.*

The refrigerator has been at the centre of civilisation ever since. Even the American NSA whistleblower Edward Snowden is reported to have obliged visitors to put their mobile phones in the fridge – at least in the early days of his still-ongoing story. One assumes Snowden wanted to shield the phones from the outside world and cut off any attempts by forces known and unknown to use them to eavesdrop on conversations. The thick insulation would certainly muffle voices, but it has been suggested that the fridge was being used as a Faraday cage, a space that is shielded from electromagnetic radiation, like radio waves used to control mobile phones.

Standard fridges are not Faraday cages – your phone rings just the same inside (a cocktail shaker, on the other hand, is entirely effective in shielding the device. No doubt that is what James Bond would use).

As we shall see, the refrigerator and the use of cold is entwined in the everyday and the extraordinary. It lies behind a wealth of other modern technologies from artificial fabrics and antibiotics to test-tube babies. Low-temperature phenomena are also pointing the way to some way-out technology of the future. Science fiction seldom mentions the refrigerator, but in science fact quantum computers and teleportation machines are all going to need one.

The story of how we got to this point spans centuries and crosses the globe, but it all began, as is the case with many things, in a hole in the ground of ancient Mesopotamia.

ONE

Old-School Cool

… which never before had any king built.

Mari Tablet, *c.* 1775 BC

The recorded history of refrigeration begins on the western bank of the Euphrates in the Sumerian city of Terqa. Lying undiscovered until 1910, Terqa is just inside what is now Syria, a few miles from the Iraqi border, in a location that has seldom been far from world-changing events. Little is known about the city until it hit the headlines in the eighteenth century BC, when it came under the control of Zimri-lim, the king of Mari, a city a few miles downstream. Zimri-lim had had his troubles, having lost his kingdom to the Assyrians in his youth, and once back on the throne he set about turning Mari and his satellite cities into the envy of Mesopotamia. The Mari Tablets, unearthed in 1933 during excavations of the capital, report that Zimri-lim ordered the construction of the Terqa icehouse 'which never before had any king built.'

The analysis of the use of ice in Mesopotamia at the time is fraught with confusion, mainly because chroniclers made no clear distinction between ice and copper ore, two highly valued substances in a Bronze Age society. Both substances are recorded as *suripum*, meaning 'fusible'. That was logical enough in the days when even the processes of melting and freezing were imbued with a certain marvel and magic: ice transforms into water and vice versa, just as ore, fired in a charcoal furnace, releases liquid, a molten metal that hardens into solid copper.

Nevertheless it appears that Zimri-lim's decree does relate to the cool of ice, not the heat of a smelter, but using ice per se was not his innovation. Indeed, the Assyrian ruler

Shamshi-Adad, who had taken charge of the region during Zimri-lim's interregnum, had already acted to secure the icy resources of his new domain. He ordered his son, who sat on the throne of Mari on his behalf, to collect ice from a location many miles (between 20 and 40 is the estimate) from the city. He gave instructions for it to be guarded day and night, and to have it washed 'free of twigs and dung and dirt' before the cup-bearers used it to prepare cooling drinks.

Letters from contemporaries indicate that there was at least one other extant ice store at the time, to the east at Qatara. The people of Ur, far to the south, were also using ice to cool their wine – although it appears that the Mesopotamian custom was to drink a beer warm – and it is assumed that Ur and other cities must also have had ice stores. However, it is conjectured that these were merely pits, timber-lined holes in the ground, where ice collected from the mountains was buried to keep it cool and secure.

Zimri-lim's construction at Terqa, although never found, is thought to have been much more elaborate. The instructions state that it was six by twelve metres and had channels to remove any melted water and so prolong the life of the ice in the Middle Eastern heat. It has been suggested that the Terqa icehouse – and a second commissioned at Saggaratum further north – was designed to make ice in situ using a freezing pool. The occasional clear winter night in Zimri-lim's kingdom may well have been cold enough to freeze shallow water, but the earliest solid evidence for this ice-production system comes from ancient Persia, where a colder climate coupled with a marvellous cooling technology was developed in the fifth century BC.

Whatever the precise function of Zimri-lim's icehouses, they were certainly at the limits of construction abilities at the time. The second one at Saggaratum was delayed by a long search to find wooden beams that were as long and sturdy as the design demanded. Time was not on Zimri-lim's side, however. In 1761 BC, or thereabouts, he fell foul of his Babylonian allies, who more or less wiped his kingdom – and its icehouses – from the face of the Earth.

❋

Further south, in ancient Egypt, ice was in very short supply. Ice-capped mountains were many days journey away, and the only natural ice most Egyptians would have seen arrived as violent hailstorms – sent on behalf of Moses or otherwise. Desert storms are remarkably violent – in 2010, hail killed four people in Egypt and injured dozens more – but any ice they provide disappears almost as quickly as it arrives. Therefore, the pharaoh and his court relied on a different cooling technology to chill their wine.

Tomb paintings dating from around 1400 BC show slaves hard at work fanning racks of jars. This has been interpreted as a means of chilling the wine within. We know that Egyptian wine was stored in earthenware – as it was in the Greek and Roman eras to come. We also know that these jars, perhaps better understood as amphorae, although that is a Roman term, were stowed away in underground cellars insulated with mud. Down there the wine was shielded from the heat of the baking sun up top, but would have been decidedly warm nevertheless.

Athenaeus, a Hellenised Egyptian writing in the third century AD, records how his ancient forebears set about making the wine cooler. Slaves would haul an amphora or two out of the cellars as night fell and place it on the highest part of the roof, where the cool night breezes were strongest. Each amphora was placed in a shallow water bath, and the slaves were tasked with sprinkling the jar with water, ensuring that it stayed wet all night long. Earthenware is quite porous, so the jar would absorb much of the water on its surface. As the night wind whistled past, the trapped water would evaporate, and the wine in the jar would get – ever so gradually – colder. If the wind was insufficient, the slaves would have to create an airflow with fans, and these are the labours recorded for posterity in tomb paintings. For an Egyptian king, all it took for a drink of cool wine was a single night of frequent sprinkling and frantic fanning by a team of indentured servants.

The busy slaves, and their drunk pharaoh, had no inkling of the physical process. The kernel of the idea probably came from a breeze feeling cool on the skin, and positively chilly on skin dripping with sweat. So in effect they were simply making the wine jar 'sweat' too. Many Enlightenment minds would wrestle with the idea centuries later and reveal that transforming liquid water into a gas, or vapour, requires an input of energy. So, with every drop of water carried away by the wind, a little bit of heat energy was taken away as well.

Incidentally, the same process of evaporative cooling is the purpose of the pith helmet, now an icon of nineteenth-century imperialist oppression as worn by irascible Europeans. Fittingly, the pith helmet was designed to chill out a colonial hot-head; although it looked rather heavy, the helmet had no protective function. Instead it had a frame made from lightweight cork-like pith, the spongy interior of certain plant stems. White fabric covered the frame to create shade. The fabric also had holes in it to allow air to circulate through the space above the head. A thick band of folded muslin, generally called the *puggaree*, the Hindi word for turban, was wrapped around the base of the hat. For a fully functioning pith helmet this fabric was kept wet, which in turn saturated the fabric cover, and created an evaporative cooler no different from the rooftop wine chillers of ancient Egypt.

Using earthenware as evaporative coolers is also recorded in ancient India, perhaps an example of a technology transfer along ancient trade routes, although the direction of travel is unclear. There the cold winds and lower night-time temperatures made it possible to cool the contents of a jar – generally water this time – until they froze. Freshly boiled water was used to charge the vessels, because it froze more quickly than cold water. Why this should be so is still debated by physicists today – and bemoaned by motorists who pour hot water from a kettle onto a frozen windscreen during Arctic blasts, only to find that it freezes up again all too rapidly. The likes of Aristotle and René Descartes had stabs at explaining it (to no avail), and today it is known as the Mpemba Effect, after Erasto Mpemba, a Tanzanian who put the phenomenon into a scientific context while still a schoolboy in 1963.

❄

Indian communities also manufactured ice using shallow freezing ponds, as has been conjectured to be the primary purpose of Zimri-lim's primordial icehouse. Whatever the truth of that assertion, it is certain that by the start of the fifth century BC, the plateaux of Persia to the east of Mesopotamia had become the centre of ice technology. Aspects of that technology have long since spread east and west – and were even exported to Latin America in the sixteenth century.

Persian cooling technology was based on three components, the *badgir*, *qanat* and *yakhchal*. The *yakhchal* literally means 'ice pit', and in modern Farsi it remains the word for refrigerator. A *qanat* is a type of underground irrigation channel, while *badgir* means 'windcatcher', and is a simple but effective form of air conditioning.

The reasons why Persia became the ice capital of the ancient world are twofold. Firstly, the climate has a wide temperature range. Winter nights are frequently cold enough to freeze water, and the days are warm enough in summer to make ice a valued commodity. Secondly, the region is dry without any large rivers. The low humidity in the air makes it easier for ice to form, and the lack of surface water necessitated an irrigation technology that tapped into groundwater, but kept it underground to prevent evaporation. This in turn led to a mastery of water management.

The people of Iran and Greater Persia have relied on *qanats* to supply fresh water for homes and fields for the best part of three millennia. They were cut by hand into rocky hillsides to access the groundwater deep inside. The work was carried out by skilled tunnellers known as muqannis. Their dangerous profession was essential to almost every settlement and this was reflected in their huge pay packets – and the jealousy with which they guarded their working practices.

A *qanat* was constructed by a team of three or four muqannis, generally members of one family. The first step was to find a source of water. The experts would survey a hillside looking for natural springs and seasonal streams and then search on the hillside above, especially around areas of thick vegetation, for indications of where to access the water table underground. Next they would sink tester wells until they found a suitable source of water. Then the construction could begin.

The concept of a *qanat* is a gently sloping channel down which fresh water flows, passing under the hillside and emerging onto the plain beneath. Digging began at the water's ultimate destination, most often fields that needed irrigating. The normal process involved sinking a number of shafts between that start point and the source well. This sped up the process by making it easier to remove spoil from each stage of the excavations, rather than having to haul it all the way back to the start. The excavated soil and rock was piled up around the shaft openings. This not only strengthened the shafts and prevented debris from being washed in during rain, but also made it easier for people out on the hillside to spot the potentially deadly drops.

It was not uncommon for *qanats* to be several hundred metres long, with many being longer. One serving the city of Kerman, situated at the edge of the Lut Desert, was 70km (45 miles). These longer qanats were an entire underground canal system, with tributary channels joining the main artery to boost water levels. It goes without saying that such a project would have taken decades to complete.

The gradient of the *qanat* was crucial. If it were too shallow, the water would not flow in the required quantities. Too steep and the water would form a subterranean torrent that would erode away at the tunnel walls, leading to their eventual collapse. *Muqannis* would install fired-clay linings to strengthen muddy sections of the tunnel, and even construct artificial waterfalls where steep gradients required it.

With the fields being the ultimate destination of the water, settlements were constructed upstream. The wealthier families had homes located uphill from the *qanat*'s outlet. Once the water entered open-air channels it became prone to contamination, and ever more so as it gushed past successive neighbourhoods. The very poorest communities had to make do with water that had made this journey to the bottom end of town, and in summer when the water table was low, they were lucky to receive much more than a trickle. Meanwhile, the wells in the upper section of the town pulled water straight from the qanat while it was still underground and not yet out into the open. The wealthiest of all had a private reservoir connected to the system, and a *qanat*-connected des-res would have not only clean water, but also the means to cool its interior rooms. For this, the house would also need a *badgir*, although the translation of windcatcher does not quite fit in this case.

The Persian *badgir* was not the first device to use of the wind to cool a building. The Egyptian windcatcher design, *malqaf* in Arabic, dates from the fourteenth century BC, around the same time those slaves were cooling wine on the roof. It was a tall, chimney-like tower with one open side, or perhaps simply a series of openings high up on the side of a building, always facing into the prevailing wind. Its purpose was to scoop in a flow of air and direct it down into the body of the dwelling. When used in conjunction with outlets on the other side of the building, the windcatcher created a supply of cool, fresh breeze that blew away the hot, stale air that would otherwise be trapped inside.

This kind of windcatcher was one of the architectural tools used across the torrid zones of the ancient world – along with open courtyards, tall walls and small, high windows – to maximise the shade. However, a windcatcher that relies on the cooling effect of a direct flow of air is only truly effective if it collects a breeze colder than the air already inside the house. When the prevailing wind becomes a hot desert blast laden with dust and sand, the windcatcher becomes more of a hindrance than a help.

The Persian *badgir* consisted of an altogether more sophisticated system. It too employed a chimney-like tower, but this time the outlet faced away from the prevailing winds. Shutters could be opened and closed when the wind changed direction. As the wind whistled around the tower it had the effect of drawing out air from inside the tower. This is an example of the Coanda Effect,* which explains that a flow of air or other fluid has a tendency to be attracted to solid surfaces, in this case the outside of the tower. This creates a pressure drop as the wind moves past the tower, which is then equalised by air from inside rushing out.

The ancient *badgir* had no need for jet power. The flow of air up the tower pulled warm air out of the house below, most notably from a basement room, or *shabestan*, which became the main living quarters in the hottest part of the summer.

*This particular eponym is for Henri Coanda, a Romanian engineer who is hailed as the inventor of the jet engine in his home country. His turbine-powered aircraft was built some 20 years before the English inventor Frank Whittle filed his more widely accepted patent in 1930. However, the Hungarian aircraft never got off the ground, with its first and only journey ending in a fireball on the runway. Nevertheless, Coanda did succeed in building Cyril Vladimirovich, the Grand Duke of Russia, a jet-powered snow sled, which is reported, rather dubiously, to have reached just shy of 100km (40 miles) per hour.

The *shabestan* was connected to the *qanat* through a shaft in the floor, through which fresh air entered the room. This supply of air had in turn been sucked into the system through the service shafts further upstream. Originating outside in the direct heat of the sun, this air was hot when it flowed down into the *qanat*, but was chilled by the flow of water, and this cooling was transferred to the *shabestan*.

The *badgir* was not used purely for keeping wealthy Persians cool in summer. Larger *qanat* systems would become overloaded in winter, when more water was entering the system, or overnight, when less was being taken out. Therefore excess water was diverted by underground canals to domed cisterns, known as ab anbars (literally 'water store') where it could be kept until summer. Badgirs were included on many ab anbars, and were highly effective, keeping the water inside just above freezing.

Similar air-cooled stores would be used to store snow harvested from the mountains, but the *yakhchal*, or ice pit, took the process to an entirely new level. Despite the name, the *yakhchal* was both an ice factory and an ice store. The main components were a domed store with massive walls made from a specially mixed mortar and a long, tall wall that ran east to west. Ice was made in winter by flooding the ground on the north side of the wall with cold water from the *qanat*. The wall would cast a shadow over the pool for most of the day, ensuring that it did not evaporate away in the sun and keeping it at the same chilled temperature at which it left the underground supply. A good freezing night was preceded by a clear, cloudless day. As night fell, any heat that had been absorbed by the ground by day radiated back into the now sunless sky. Low clouds blocked the escape of heat, but on a starry night ground temperatures dropped rapidly, and by dawn the shallow pool was frozen solid. The process of getting it into the store would already be underway, with the ice being smashed into pieces and carried through a tunnel in the thick base wall of the icehouse.

The inside space of a *yakhchal* itself was a little deceptive. From the outside most looked like conical domes, but the ice is not stored inside here. Instead the dome was a cover, a kind of clay pith helmet, over a huge pit dug into the ground. Below ground was deemed to be always cooler than above, and this is where the ice was packed, sealed by a layer of straw and sawdust to further insulate it. The internal walls of the building above ground thicken towards the base, so what looks like a dome on the outside is more of a chimney when seen from inside, although still distinctly conical. In effect, the dome was part of a giant *badgir*.

Small holes were cut into the apex of the dome to release the warm air rising up from inside as air cooled by a *qanat* took its place lower down, using the same system as cisterns and shabestans. In some cases, four wind-catching towers were positioned around the dome, and together they were able to catch the breeze from any direction. This stream of air was not cooled in itself, but as it was forced up inside the conical interior space, it was squeezed into the decreasing space available. At the top, this compressed air whistled out of the vents at great speed, taking any residual heat with it.

While earlier cold stores used timber or mud linings, a *yakhchal* was built from a material called *sarooj*. This was a mortar made from sand, lime and clay, as you might expect, but also egg whites, goat hair and ash. The combination was not only waterproof but also an excellent insulator, keeping the heat out and the cold in. *Sarooj* was used both to make the dome – at the base the walls were around two metres (6½ feet) thick – and also to line the pit hidden beneath. Any meltwater that trickled out of the ice was collected and either added to the cooling system or recycled for the next ice crop.

Some yakhchals were smooth on the outside, while others had a stepped appearance. It has been suggested that these steps were used to hold a loose thatching of straw as added insulation from direct sunlight. The straw was put in place during the day, but removed in the evening, so the dome could radiate any residual heat back into the night sky.

❄

Producing ice on this industrial scale made it a mass-market commodity. Blocks of ice were extracted from the store in summer and taken by donkey to markets all over the region. Ordinary people, not just the kings and courtiers, would have been able to enjoy the delights of ice puddings and cold drinks. To this day one of the favourite puddings in Iran is *faloodeh*, a sweet confection of rice noodles, nuts and chilled milk. However, there is little evidence that ice was purchased to preserve foods on any great scale. The *yakchal* itself would have acted as a cold store for perishable goods, probably fruits and fresh meat, but these would have been the preserve (literally and figuratively) of the wealthy few.

Ancient cuisines were not entirely focused on the next meal as they are today. The preparation of food was a year-round affair. Cooking not only makes food safe to eat, but it also breaks down the chemical constituents making it easier for us to digest. It is claimed that a modern human is unable to produce the right mix of stomach enzymes to handle raw foods, and could not survive for long without being able to prepare foods in some way.

Excess food during time of plenty has to be preserved, and cooking it is not enough. Root vegetables can be simply buried. If packed carefully in cold and, above all, dry ground they will survive untainted for months. If any damp creeps in, however, rot sets in within days. Drying is a more secure method of preservation, and probably the most ancient. It removes the water from the food and kills any of the bugs growing on it. When there is not enough sun to dry the food, smoke from a fire has a similar effect.

Salting is another fast and effective way of drying out food. It works by drawing the water out of the food – and the germs – by the process of osmosis. Briefly, this is a phenomenon that occurs when liquids are separated by a membrane. In the case of meat and produce, the liquid is water mixed with a range of other substances and the membrane is the one that surrounds all living cells. Such a membrane is semi-permeable. This means that water can pass through it but most other substances cannot. The molecules in a liquid are in constant motion in random directions, so they naturally spread out to fill a space, until blocked by an impassable surface. During osmosis, the water can spread out through the membrane but the other materials cannot. If the concentration of the water – the amount of other things mixed into it – is equal on both sides of the membrane, the water will move in both directions in equal amounts and there is no outward effect. If the concentration is higher inside the cell, water from outside floods in to equalise the concentrations. When there is a large difference, so much water enters the cell that it bursts. This is why meat goes floppy and grey when left in fresh water: osmosis has ruptured the cells, damaging the structural integrity of the flesh and releasing all its juices. Salting the surface of the meat creates a higher concentration outside the cells, so water floods out of the food, and – crucially – any germs that populate it. Washing away the salt again allows the food to be partially rehydrated.

Surrounding food with high concentrations of sugar works in the same way, and it is a sugary preserve that has been the lasting legacy of the Persian ice culture. *Sharbat* is an iced treat that hails from the region. The name derives from the Persian word for 'drink'. It seems that in ancient Persia – at least from the earliest records from the second century AD – a drink without the benefit of ice was unthinkable. In its original form *sharbat* is a sweetened syrup cordial, flavoured with fruits, spices or flower oils (the word syrup has the same root as *sharbat*). The syrup can be diluted with ice water to make a refreshing drink, or mixed with crushed ice to make a dessert. *Sharbat* is still a popular drink across the Muslim world. In a throwback to the Ottoman Empire it is traditionally sold by vendors who carry the chilled drink in large flasks on their backs. They pour it into a glass for each customer from a spout that curves over the shoulder, bowing forward with a flourish as they do so. Such a scene is mainly for the tourists nowadays, although refreshing sharbats are still sold in cafes and fast-food restaurants alongside the more familiar bottles of carbonated drinks.

Wine, beer and other alcoholic drinks are not allowed by Islamic law. These types of drinks arose not purely due to their intoxicating qualities but because the alcohol killed most of the germs in them. Thus in the absence of clean drinking water, imbibing a cup of beer or wine was the safest way to hydrate yourself – although far from ideal. Before the advent of good sanitation, European urbanites were constantly tipsy. By contrast, the Persian *qanat* technology, which spread with the Islamic Empire, supplied a clean product, and the sugar content of the syrup was high enough to stop putrefaction. Therefore, *sharbat* was an excellent – in fact much better – alternative for quenching thirst than brewed beverages.

Persian ice technology took centuries to perfect. Many of the historic yakhchals seen today in Iran and the surrounding region are relatively recent constructions, and were used well into the twentieth century. Without the investment in building *qanat* networks, or the need to, the Persian technology had less of an impact on other parts of the world. It was used mainly to store snow harvested in winter, but not to make ice.

Alexander the Great's army returned from his conquests of Persia and beyond in the fourth century BC with the basic know-how to build a functioning snow store, insulating a pit with straw and cut branches. The Greeks (and Romans after them) used stored snow to cool drinks, either mixing it directly with wine, or using an early form of wine cooler called a *psykter*. This was a type of ceramic pot that dates from at least the sixth century BC. It had a rounded body standing on a tall, sturdy foot. A *psykter* was placed inside a larger vessel called a *krater* (basically a large mixing bowl). The *krater* was filled with snow so it surrounded the *psykter*, which was kept filled with wine. The result was the ice bucket of the Classical era, although it should be noted that the paired vessels were used for warming drinks as well. An alternative set-up has been suggested where the snow was in the *pyskter* and that was plunged into a *krater* of wine. One assumes it came down to how much wine and how much snow one had.

The iced dessert form of *sharbat* is now known in the West as sorbet, a smooth mix of fruit and ice, having been adopted and perfected by Italian chefs over the centuries. The story goes that Nero, the mad Roman emperor, began the fashion for sorbet by eating honeyed fruits chilled with snow rushed in from distant mountains. This is probably a bending of the truth, meant to highlight Nero's megalomania. Roman nobility – and not all of it psychopathic – was eating desserts chilled with ice long before this. However, there was a practice of sweetening foods with a compound of vinegar and lead, which no doubt had an impact on your average Roman's faculties, mental as well as physical.

In American English, the word sorbet refers to the same dairy-free iced pudding, while sherbet (another derivation of *sharbat*) is a kind of low-dairy ice cream. In British English, sherbet refers to something very different, a sickly sweet powder mixed with carbonates. Launched in the early 1800s, sherbet was originally meant to be added to water to make a sparkling fruit drink. Today, however, it is consumed dry, and when mixed with saliva, the powder fizzes up in the mouth, thought by some to be a pleasant outcome. This confection is more like a third form of *sharbat* from Central

Asia, where it is served as a solid sweet (although not fizzy) and mixed with honey and nuts.

❄

So far we have ignored the contribution of Chinese technology and other Far Eastern cultures. Poetry from the eleventh century BC makes reference to ice being cut from frozen lakes and stored until the summer, and cellars stocked with ice are reported to have been used to store meat during the Han dynasty (second century BC to second century AD). In the fifth century, emissaries from northern China (on behalf of the Tuoba Wei dynasty) made several visits to Persia. They returned to tell of many marvels, not least the *qanat* waterways, the plentiful ice in desert cities and the homes kept cool in summer. The Chinese must have been impressed, for by the Tang dynasty of the seventh century blocks of ice were being used to cool the homes of the wealthy, and the emperor's staff included 94 workers under the auspices of a court iceman to collect and maintain stores of snow.

As well as chilling wine and drinks, records show that Chinese icehouses were used to keep fruits and vegetables. Melons seem to have taken up the most room. By the seventeenth century, during the Ming dynasty, perishable foods – everything from summer fruits and bamboo shoots to fish – were being transported long distances on barges refrigerated with blocks of ice, cut from the mountain lakes upstream the previous winter.

A Japanese emperor, Nintoku, who ruled in the fourth century AD, was also so enamoured of ice, after being given some by his son (some reports say it was a brother), that he declared the first of June every year to be the Day of Ice. On that day the work of government ceased and officials and military men came to the imperial palace to receive a gift of ice, so they too could enjoy its wonders.

The Joseon dynasty, which dominated Korea from 1392 to 1897, was also using ice for preservation, but in a rather more macabre way. By 1396 the capital at Han-Yang (now Seoul) had two icehouses: Dongbinggo, the eastern warehouse, and Seobinggo, in the west of the city, were stocked with ice cut when the Han River froze each winter. The ice in Seobinggo was used for culinary purposes, most notably for *bingsu*. Still a popular Korean pudding today, this dessert is made of ice shavings with a sweetened topping. The original Joseon version would have had a dollop of sweetened red-bean porridge.

The distribution of ice from the royal stores was based on rank, and this was especially important for the contents of Dongbinggo, where the ice was used for rituals, not victuals. Joseon society adhered to Confucian principles, and this included a great respect for the dead. Funeral rites were observed down to the most minute detail, and included a period when the body, dressed in fine clothes, was watched over by mourners prior to burial. Status was all, and this was reflected in the length of mourning period. When a king died the burial would not occur for some time, as tradition demanded that his body lie in state for as long as possible.

Whenever a king or queen became seriously ill, a state of emergency was declared, and the icemen at Dongbinggo went on full alert. If the worst happened, they would be ready. The royal corpse, dressed in white silk, would lie in state in an open coffin for five months. To stop nature taking its course, the coffin was placed on a platform made of several ice slabs, held inside a bamboo frame. Meltwater was soaked up by dried seaweed packed around the ice. The ice and seaweed had to be replenished day and night. Dongbinggo held 15,000 blocks of ice, each cut to the approximate length of a coffin. If the king died in spring, Dongbinggo would be empty by the time of his burial in the autumn.

Dongbinggo and Seobinggo are now districts of Seoul, near the north bank of the river. The icehouses closed in

1898 but the buildings are still intact, and are surrounded by cafes that serve up *bingsu* in the traditional way.*

❋

But what of ice cream, that mainstay of the modern fridge-freezer? Unsurprisingly it plays a role in the development of the refrigerator. The earliest foodstuff that could be described as ice cream was developed in China. The Tang dynasty, clearly aficionados of cool, made a frozen dessert from fermented buffalo milk, flour and camphor. Camphor, for those not in the know, is more commonly associated with paint stripper, cough mixtures and embalming fluid, but is still used to flavour sweets in Asia. It is thought that the waxy qualities of camphor would have given the dessert a flaky texture, and eating it would have been similar to eating snowflakes. The tangy Tang ice cream was made in metal tubes that were packed in ice, in the same way that *kulfi*, Indian ice pudding, is made today.

It was an acquired taste, no doubt, but this is probably where ice cream starts. However, a legend tells how the Chinese court learned of the dish from Mongolian horsemen, who came across it by accident. The Mongol warrior lived life in the saddle, and carried food with him inside leather bags made from the intestinal tracts of horses and goats and other large herbivores that roamed the steppe. The story goes that when the warriors filled these

*Refrigeration has also transformed the preparation of Korea's national dish, kimchi, a wonderfully varied concoction of fermented vegetables. Traditionally it was prepared by burying ingredients in a jar, which ensured the cool but frost-free conditions needed to slowly create the distinctive acidic tastes adored by Koreans – but met with puzzlement further afield. In 1995, kimchi-making entered the modern era with the invention of the kimchi refrigerator. This is not simply a chiller; it keeps the cold air within still and humid, thus emulating a hole in the ground. The fridge even tackles kimchi's strong smell, neutralising the whiffs using powerful LEDs. Today, four out of five South Korean households have a kimchi fridge.

bags with cream (probably from a mare) and rode through freezing winter conditions, they found at journey's end that the bags were filled with a light and fluffy mixture. The combined action of the cold and the constant churning transformed the cream into ice cream, which is a complex mixture of tiny pieces of solid milk fat and ice (plus, one hopes, some dissolved sugar and flavourings).

Perhaps this was how ice cream came to be, although it should be noted that similar stories also attribute the invention of the burger and cheese to nomadic tribesmen. The first cheese is conjectured to have been made when goat's milk was stored in a bag newly made from an animal stomach. The residual enzymes in the stomach lining coagulated the milk into cheesy lumps. The burger, meanwhile, came about when meat stored under a saddle was minced by the weight of the rider on top – and all the while slow-cooked by the body heat of the horse underneath. A mostly raw version eventually became something of a delicacy in Europe as steak tartare, named after the Tartars, a western band of the Mongol horde, hailing from what is currently southern Russia. Following the story to its conclusion, steak tartare spread from Russia through the Baltic trade routes until it arrived on the docksides of New York, sold as Hamburg steak to new arrivals. It would appear that even a cheeseburger and iced shake have a lot of history.

Another creation story for ice cream has its secrets being brought to Venice in the late thirteenth century by Marco Polo, returning from his epic journey of discovery through Asia (if indeed he ever left home*). Yet another sees it brought to the world stage by Catherine de' Medici, an Italian who married into the French royal family. Whatever truth these myths contain, the ice cream we enjoy today is irrefutably a consequence of science joining the quest for cold.

*Some schools of thought suggest that Polo never got beyond the prison in Genoa where he dictated his tales to Rustichello da Pisa, the actual author of famed travels.

TWO

Conjuring Cold

For heat and cold are nature's two hands, whereby she chiefly worketh.

Francis Bacon, 1624

As all good mysteries should, the story of how cold came under the purview of science begins with a poisoning. In 1536, the Dauphin François, the heir to the French throne, was returning victorious from a defeat of Charles V's troops. The Holy Roman Emperor had been attempting to annex Provence, but to no avail. After a bout of strenuous exercise (he had been playing *jeu de paume*, a precursor to tennis) François called for a cooling draught. His cup-bearer, the Italian count Sebastiano Montecuculli, duly complied with some iced water – and François collapsed. He never recovered and died within the month.

It was widely believed at the time (an age of many poisonings) that the expert poisoner preferred to disguise his or her toxicant in iced water rather than a warm drink. The victim was much more likely to gulp it down in one go, and less likely to taste it – until it was too late. Montecuculli was subsequently tortured, and he duly confessed: he had poisoned the Dauphin on the orders of Charles V, and the French king was next on the list.

Montecuculli lost his life, but there was no love lost between the two embattled states. Charles V had already imprisoned François and his younger brother Henri in a Madrid dungeon for three years, when they were still boys, and François had never really got over the tuberculosis he contracted while incarcerated. This deadly illness seems to have been somewhat overlooked during the inquest into his death. The sons of kings are never weaklings, after all. A death so young (François was 22) could only be by murder.

Brother Henri now became the Dauphin and his wife Catherine de' Medici, the niece of the incumbent Pope, was set to one day be the queen of France. The French people did not like the idea of a Florentine sitting on their throne, and one from a nouveau-riche banking family to boot. The story goes that their suspicions and distaste for Catherine were inflamed by how she had wheedled her way into the royal court's affections. Still a girl of 14 when she married Henri in 1533, she had arrived in France, so people were told, with an entourage of Italian chefs. On her orders they offered up all wonders of culinary feats, not least a confection of ice, cream and fruit – ice cream. So the young girl who had debased the royal court with her decadent iced fancies (and had introduced forks to the table!) was now to be queen, and all thanks to another perfidious Italian bearing ice. You couldn't make it up.

In fact, you could. The story of the Medici girl bringing ice cream to France is the truest confection. It appears to have been a conclusion arrived at in the nineteenth century, and despite no evidence for it and plenty against, the tale has become part of the brand story of Italian gelato, repeatedly retold by the self-proclaimed makers of the world's best ice cream.

The legend does have a tangential link with historical fact. It was not Catherine but her son, Henri III, who brought the craze for iced treats (and forks) to France in the late 1580s. And the recipes did indeed come from Italy (although they were probably for sorbets and iced wines, not true ice cream). It is likely that the still-hated Catherine was held responsible for the king's love of this foreign muck, and over the years that blame became conflated with aspects of her earlier history.

By the late 1580s, iced food was indeed the height of fashion among the noble classes of Italy. It is said that Francesco de'Medici, the Grand Duke of Tuscany, had got himself addicted to an alcoholic ice cream-like drink made from eggnog, which he prepared himself. Francesco was something of a secret alchemist, who spent many a night in a laboratory in the Palazzo Vecchio in Florence. He had long been an advocate, although again out of public view, of the use of harvested snow and ice for medicinal purposes. As in France, ice was still held with suspicion by the masses and the Inquisition, who saw it as having supernatural qualities, the kind of thing a witch or wizard might use, or an alchemist, which was much the same thing.

It is entirely possible that Francesco had heard of – and was using – a mysterious new technique for making ice that was being perfected by Giambattista della Porta in Naples. The Neopolitan polymath, also known as 'the professor of secrets', was the nearest thing Italy had to a wizard at the time. He had already been carpeted by the Spanish Inquisition (Naples was a Spanish possession in those days). The cardinals gave him a talking to about not unearthing too many secrets of nature. Giambattista avoided imprisonment, although some of his colleagues did not. He compounded his reputation by communicating with them in prison by writing messages on the insides of boiled eggs, then given as innocuous gifts. Something of a showman, he continued to make spectacular public performances. One of these involved freezing bottles of diluted wine by swirling them in a fantastical magical mixture.

He published his techniques in the 1589 edition of his aptly titled work *Magia Naturalis*. Until this point, no one had really questioned that magic was an integral part of natural phenomena. But within 30 years the grand dining tables of Europe were serving ices made using the very same occult method. Could something that was becoming so ordinary be all that magical? Natural philosophers wanted to find out.

❋

By the end of the sixteenth century, the understanding of hot and cold had moved little from the explanation set out in Classical Greece. The world was believed to be composed entirely of earth, water, fire and air. Empedocles, whose poetic account *On Nature*, dating from the fifth century BC, is the earliest surviving reference to this theory, puts it another way:

> *Hear first the four roots of all things: Bright Zeus, life-giving Hera and Aidoneus, and Nestis who moistens the springs of men with her tears.*

Zeus, the ruler of the gods, represents the Sun and its fire, Hera, Zeus's wife, represents the sky and its air, while Aidoneus is another name for Hades, the god of the underworld, who resides deep among rocks and earth. Nestis is another name for Persephone, the wife of Hades, who spends half the year with him and the other half up top with her mother Demeter, the goddess of the harvest. Persephone and the gurgling spring waters she represented bridged the gap between the underworld and the world under the sky.

Empedocles had not just conjured this idea from the heavens. He had performed a scientific experiment, one of the first recorded. However, in the modern context it is the kind of thing three-year-olds discover in the bath. Empedocles used a clepsydra, a kind of long pipette or 'water thief' that picked up water from a deep barrel. It had a hole at both ends and a vessel at the bottom. You dipped it into liquid; water flooded in the lower hole, and you covered the upper hole with a thumb or fingertip as you raised the device up, thus keeping the liquid trapped inside. (Covering the top of a drinking straw does the same thing – watch the barman do it the next time you are out for cocktails.) Empedocles covered the top hole of an empty clepsydra before lowering the whole thing into water. He found that no water (or not much) entered through the lower hole, until he released his finger from the top. He took this as evidence that the air, although invisible, was made of material.

From this he built a central thesis that the content of the physical world was derived from the four elements. It was finite and constant, but it was also in perpetual change. He believed that disorder and strife drove the elements apart, while love and harmony pulled them together in different proportions to make up the body of nature. If ever love conquered strife, or vice versa, life would cease to exist.

But then again, Empedocles believed a lot of things. He lived in Agrigentum, a Greek colony on the island of Sicily. He subscribed to many of the teachings of Pythagoras, who had lived over on the Italian mainland a generation before him and had believed in the transmigration of the soul on death. Those who did wrong in human form came back as animals – Empedocles was therefore a vegetarian – and those who accrued wisdom escaped the cycle of rebirth.

And to prove it Empedocles threw himself into a volcano, Sicily's mighty Mount Etna. His thinking is unclear, but one suggestion was that he was aiming, in his wisdom, to become an immortal fire god.*

One thing Empedocles did not do was use the word 'element'. That came around a century later with the work of Plato. However, Plato was an idealist, meaning he thought that truth lay beyond the senses, and so formulated an intellectual framework that explained how reality worked. For him nature was divided into two realms, the realm of senses and the realm of *eidos* (the root of the word 'idea', but better translated as 'form'). Only forms were truly real, but our senses could detect mere shadows of them. Thus life was simply (or perhaps not) an illusion.

According to Plato, the forms of the four elements were four of the Platonic solids. The Platonic solids are the only

*He was almost right. In 2006 an undersea volcano found near Sicily was named after him.

regular polyhedra possible in three dimensions. They are shapes with edges that are all of equal length, making faces of equal area. There are five of them, the most obvious being the cube. For Plato the cube was the form of earth, the tetrahedron (the simplest shape) was fire, the octrahedron was air and the icosahedron (made from 20 triangles) was water. To Plato, the fifth solid, the complex dodecahedron made up of pentagons, was the nearest to a sphere, and he opted for this to be the shape of the Universe itself.

Our word 'element' is based on the Latin for 'rudiment', but the classical concept of elements was not confined to the ancient Greek or by extension the Western world-view. Similar ideas were found in India and China (although they sometimes opted for five or six elements, adding in metal and wood). Like these Eastern traditions, the Greek concept of elements was not the basis for just physical substances; they were also bound up with gender, emotion, good and evil, and of course hot and cold.

Disease was ascribed to an imbalance in elements in the body. A fever, for example, was the consequence of too much fire, and a runny nose of too much water. Each element was represented by a corresponding humour, the physical manifestation of a temperament. A sanguine attitude was due to an airy optimism, phlegmatics were calmed by still waters (and a lot of phlegm), a quick temper was fuelled by fire-filled bile, while the melancholy were weighed down by a dark and earthy constitution.

Aristotle, Plato's star pupil (at least in hindsight) rejected his teacher's theory of forms (for example, the smarty pants wanted to know what the form of *forms* was). Instead he became the first true empiricist, seeking to understand the natural order by what he could observe it doing. To the modern mind we are now on safer ground, and it is Aristotle's physics that would have formed the basis for della Porta's discussions on 'nature's secrets' with his alchemist friends – and with the Inquisition.

Despite it being simplistic (and more or less completely wrong), Aristotle's description of the elements, and how they interact to create natural phenomena, is pleasingly cohesive. It is perhaps due to this harmony that it was assumed to be correct for at least 1,800 years (the Catholic Church adding this 'received' wisdom from the days before Christ to its dogma – at least some of it – also made it hard to refute).

According to Aristotle, then, every substance at hand was made from an intricate combination of the four basic materials. Fire and earth made things dry, water and air made them moist. Heavy materials were dominated by earth, while lightweight objects were permeated with air. Heat was due to the presence of fire and air (to a lesser extent), while coldness came from earth and water. The four elements also had a spiritual dimension, an aim or purpose that underwrote their actions. They were therefore nothing less than the force behind nature's ever-changing face.

Aristotle declared that all natural processes were the result of mixed elements seeking purity. The mortal realm was formed of layers of steadily purifying elements, with the heaviest – earth – forming the rocks under our feet. The oceans of water were the next layer, then the air above us and finally a sphere of fire – somewhere near the Moon – separated the world of men from the perfection of the heavens. Out there, the Universe was constructed of the fifth element, or quintessence. This unchanging and perfect 'aether' could never mix with the lowlier four.

Back on Earth nature was anything but static, as the elements sought to free themselves of each other. In Aristotle's view erupting lava was fire and water escaping from earth, rain was water separating from air and finding its natural level. The sparks from a flint were fragments of fire being released, smoke from burning wood was rising air, while the oils and resins fizzing out were its water. All that remained was the solid earthy ash. Snow and ice, meanwhile, were waters super-chilled with an earthy component that made water settle on the ground, down where it belonged.

In this way the elements were everywhere, and the idea of nothingness, or a vacuum, was impossible. Any space empty of substance was filled in the instant that it formed. In the words of Aristotle, 'Nature abhors a vacuum.' This aphorism would haunt the centuries, until it too would become embroiled in the hunt to understand cold.

For Aristotle the *primum frigidum*, the ultimate source of cold, was water. It was commonly believed that the Earth possessed a great reservoir of cold around the archipelago of Thule. This mythical land was said to be visited by the Greek explorer Pytheas, probably around the time of Aristotle's death, although no direct account remains of the voyage. On the way Pytheas claims to have discovered the British Isles (although they were known before his trip) and then visited Thule, a land six days sailing from the north of Britain. There he saw that elements appeared to converge into a world of slush and freezing mist. He also witnessed something akin to the midnight sun when the locals took him to a vista where night barely fell.

Obviously Pytheas's account chimes with a trip to the Arctic, or somewhere near to it. Educated guessers have suggested that he had skirted the North Sea and arrived on the Norwegian coast somewhere near Trondheim, before sailing home along the 'British' coast. In fact, he probably followed the northern European mainland, which would account for his skewed reports of the dimensions of Britain.

It would be unjust to say that Giambattista della Porta had unlocked the secrets of the *primum frigidum* and brought a slice of Thule to Naples. After all, according to a poem from the fourth century AD, Indian wizards had done it at least a thousand years before.

*

The first full account of the 'magical' cooling process was given in AD 1242 by Ibn Abi Usaibi, a Damascene doctor.

The idea appears quite simple on the face of it. Adding salt to water makes it get cold.

In the context of the time, a salt was a base material, a mundane matrix that could be imbued with specific qualities through a bit of hocus pocus around the cauldron (the alchemist's workshop was the forerunner of the chemist's laboratory, but is better thought of as the kind of place in which Merlin would have felt at home). Usaibi's technique worked with sodium chloride, the salty stuff we sprinkle on food, or sal ammoniac (ammonium chloride), alum (a mix of sulphates), or nitre, which is potassium nitrate, also known as saltpetre and a key component of gunpowder.

Despite the outward simplicity of Usaibi's system, it required a good deal of precision to work. The key was using salts that dissolved well in the water (some do not). As the solid crystals mix in with the water molecules, they break apart into subunits called ions. That dissolution requires an input of energy, which is taken from the surrounding water, making it colder. Adding five parts of sal ammoniac and five parts of nitre to 16 parts of water will make the temperature drop from 10°C to −12°C. But it does not necessarily freeze, instead becoming a supercooled liquid.

Of course della Porta knew nothing of the physical process taking place at the molecular level. His triumph was to figure out a method that could consistently make ice. He perfected the mixing of nitre and common salt to chill water, then added snow for good measure. He then dunked a glass vessel filled with water, or often watered-down red wine to enhance the visual effect, and gave it gentle taps and swirls to help its content solidify. The results both amazed and alarmed spectators – although some did confess to finding it all rather puerile.

The practical applications were soon being exploited, not least by Grand Duke Francesco, sozzled in his palace (in his defence it is likely that any apparent intoxication he may have suffered was also due to the fact that he was being poisoned at the time by his brother, greedy for a ducal title). However, putting murder and other practicalities aside, dabbling in 'natural magic' would surely come to no good … if it were magic, of course.

Della Porta's explanation of his cooling system was that mixing snow and nitre unleashed a chaos, and that created a 'mighty cold'. 'Chaos' was a reference to the teachings of Philip von Hohenheim, a Swiss alchemist better known as Paracelsus. If Paracelsus had still been alive, he might have suggested that della Porta had somehow accessed a secret knowledge guarded by the nature spirits, namely the gnomes and nymphs.

*

Paracelsus was a wizard of the first order – he even has a statue in Hogwarts. In the real world he died in 1541, leaving behind the most up-to-date theory of elements. Despite his prestige – his nickname was 'Bombastus' and he was exceedingly confident of his opinions – Paracelsus did not work alone. His feat was to merge the elements of Aristotle with an alternative view proposed by Jabir ibn Hayyan, remembered in Europe as Geber.

Jabir was a eighth-century Islamic scholar from eastern Persia. Like all self-respecting alchemists he needed to keep his discoveries secret, and his records are somewhat opaque in meaning. Jabir's name has been proposed as the root of the word 'gibberish'. So unclear was his writing that the main Latin translations of his work may actually be the text, at least in part, of someone entirely different.

Whether Paracelsus was inspired by Jabir himself, or by the unknown thirteenth-century translator dubbed Pseudo-Geber, is somewhat academic. The Jabirian theory was that the processes of nature were driven by two 'principles', mercury and sulphur. Slippery, fast-flowing mercury was the agent of change, while sizzling sulphur was the agent of heat and fire. Paracelsus added a third principle, salt, which was the agent of solidity, the base material.

When the four elements underwent changes they were releasing and receiving these principles. A flame was the release of sulphur, which travelled through the air as heat until it found another object to inhabit. Smoke was the release of mercury, which he described as a chaotic fluid.* The ash was the base matter, governed by the salt principle and free of the other two.

As was de rigueur for alchemists, the three principles were also tied up with unseen or occult faculties. Salt ruled the body, while sulphur governed the soul – without its fire, life would be extinguished. Gold, that ever-sought-after commodity, was frozen fire that lacked the flux of mercury and was a reservoir of the life force or 'animal spirit'. Mercury itself was a moral and intellectual force. This perhaps explains why ice, effectively water with its mercury and sulphur removed, was sometimes seen as the embodiment of evil and death. It also shows that Paracelsus saw cold as an absence of the agent of heat, rather than the presence of something else, a view seldom shared by those who came after him.

Many accused Paracelsus of dealing with demons, which were believed to be the true cause of destructive natural events such as storms, fire, earthquakes, and so on. Paracelsus did not discount the presence of non-human actors but instead presented a much more humane view of them. He said that there were four other natural beings aside from humans, each inhabiting its own element. They were not mere ciphers, however, but beings in their own right, with the capacity for emotions and morality, and able to walk among us, but generally staying out of sight.

Paracelsus admits to no confirmed sightings, but he bombastically declared that gnomes, sylphs, nymphs and salamanders were as real as 'Adamical' beings – you and me, that is. In so doing he conjured a world that could have been an alternative Narnia (albeit crucially without an ice queen).

*The word 'gas' later arrived via this link, being a Dutch adaptation of the Greek '*khaos*'.

Air is to a human, as earth is to a gnome – they breathe it and could pass through it. The gnomes were little people who lived beneath the ground, bumping into miners on occasion and making their workings collapse. Sylphs lived in air like humans, although unhindered by gravity, and were coarser, wilder versions of us, and much, much larger. These giants hung about among the treetops in woodland, and whenever they settled on the ground, an earthquake was the result. The nymphs were the water spirits and were habitually naked. As a result marriages between nymph and human were not uncommon but rarely a success. Finally, salamanders were most common near volcanoes. They were weaselly and wiry folk, unburned by fire.

Whether Giambattista della Porta was a gnome, nymph or salamander, he was not letting on. And he enjoyed the patronage of Neopolitan high society during a long life, in which he also wrote plays, designed optical devices and invented a magnetic telegraph. In the early years of the seventeenth century, another even more gnomic figure would use della Porta's know-how to send chills down the spine of a king.

❋

Cornelius Drebbel was the last of his kind. The work of the Dutch inventor, engineer and natural philosopher filled the space between spells and science. He had to work hard to maintain his magic-man image, as the powers of scientific empiricism began to take hold in Europe. Cultivating an air of mystery was Drebbel's meal ticket, as he courted the rulers of Europe with his long list of fabulous inventions.

In 1620, after a career filled with ups and downs, he enters our story inside London's Westminster Abbey in the middle of summer. He has been inside since the dawn making preparations, and is now awaiting the arrival of James I.

James was the King of Scotland who had ascended the throne of England years before to fill the gap left by

Elizabeth I, the virgin and consequently heirless monarch. James did not like the summer. Exposure to sunlight created itches and swellings on his skin. It is likely he suffered with a form of porphyria, a disease that he would have inherited from his mother, Mary, Queen of Scots, and would pass to his distant descendant George III, the 'mad' king of England. Stuck indoors on summer days, James's discomfort was compounded by the need to wear a thickly padded doublet. Since his arrival on the throne in 1603, there had been many attempts on his life – not least the much-celebrated Gunpowder Plot of 1605 – and it made sense to wear protection against the next assassin's blade.

Drebbel had served the British king some years before but was lured over to the court of Emperor Rudolph II in Prague to fill the vacancy for a castle wizard in 1610. This job landed Drebbel in a spot of bother, and he had to appeal to James to extract him from prison. In return for help, Drebbel promised to build the king an array of marvellous contraptions. Now, several years later, he was still doing his job, entertaining the king and providing him with what he wanted and needed. On this day it was a bit of air conditioning.

No records exists of how he did it, but Drebbel succeeded in chilling a side chapel in the abbey to temperatures more akin to a pleasantly brisk spring day. It is more than likely that he was using della Porta's cooling mixture, which would have been arrayed on a massive scale in tall vats along the shaded walls of the abbey. The vats may have been metal, so the cold of the liquid within was better transferred to the surrounding air. In the hours that it took to create enough mixture, the lower portion of the chapel would have become filled with cool air. Drebbel had written previously about how warm air rises, so he would have known that the warmer air would have been driven up among the ceiling vaults far above.

When the king and his entourage arrived, they were refreshed and amazed in equal measure. But the king, who sweated profusely under his quilted body armour, began to feel a bit chilly – and left, shivering. A success, of sorts.

Drebbel had come to the attention of the British royal court in 1605, a year after publishing his book *A Treatise On the Nature of the Elements* in the Netherlands. The elements of the title were not just a rerun of the classical four – this also referred to the weather, a usage of 'the elements' that persists. Specifically, Drebbel was interested in how fire, air and water were involved in creating winds and rain, and other meteors (atmospheric events). His deliberations on this subject led to one of his most famous inventions, the Perpetuum Mobile, or perpetual motion machine. Of course it was nothing of the sort, but its behaviour would be an inspiration to those who would later come to measure the quantity, or degree, of hot and cold.

Drebbel's treatise began with the four Aristotelian elements, but he was not convinced of Paracelsus's three principles. However, Drebbel did borrow the idea of putting the elements in a hierarchy from his Swiss forebear: fire was the primary element because fire could make air like fire, water like air, and earth like water.

The warmth of sunlight, therefore, created rain and wind. The heat turned water into an airlike substance, and this rose higher and higher. As it did so the airlike water was cooled by the real air around it until it turned back into liquid water and fell as rain. Similarly, wind, according to Drebbel, was a stream of warm air rushing through the sky, until it too was cooled enough to grind to a halt. He demonstrated this with an experiment. He took a retort, a teardrop-shaped glass vessel with a narrow, curved spout attached to a bulbous body, and hung it over a vat of water, so the spout was submerged. Air was trapped in the rest of the retort, and when this was heated, it began to bubble out of the spout and escape. Drebbel claimed he was making wind, although the reality of air flow would require a more nuanced understanding.

According to Drebbel's thesis, weather originated from fire, which had a heavenly origin. Elemental fire, streamed out of the Sun and into the air, where it replenished some kind of spirit, pneuma (meaning breath) or quintessence – the fifth element that not only fed flames and drove the wind and rain, but was also the vital force that created life.

Drebbel called it aerial nitre, meaning it was some kind of chaotic airlike form of nitre, one of the solid substances that would have been in his church-chilling mixture. Drebbel knew of course that nitre, or saltpetre, is also an ingredient of gunpowder, another source of fire and flame. As it happens, what nitre does in gunpowder is a bit different from its role in the cooling mixture: during the great heat of the blast, the nitre (or potassium nitrate) is giving up free oxygen, which feeds the explosion. But one can see why Drebbel was much enamoured with this miraculous substance. While he had no idea about the chemical processes occurring in either use, he had spotted a link with some 'quintessence' of nitre that was present in the air. If nitre as gunpowder made loud bangs and bright flashes, did aerial nitre make thunder and lightning? And how was it linked to the flashes of shooting stars? These are the only atmospheric events we call meteors today, but ironically they are studied by astronomers, not meteorologists.

Another two centuries would roll around before oxygen was properly identified and its role in combustion explained. However, another of the many legends associated with Drebbel tells how he managed to isolate some of this quintessence in jars by heating saltpetre. In 1621 he is reputed to have used it to replenish the air inside a submarine. Yes, he invented that as well.

The perpetuum mobile was the first thing Drebbel made when he arrived in England in 1607. It was on a much smaller scale than his other feats and to modern eyes was a lot less impressive. However, the motion of the 'captured spirit' within the contraption caused a sensation. William

Shakespeare is said to have styled Ariel, a literary enslaved spirit, in his 1611 play *The Tempest* on reports of the perpetuum mobile, and Prospero, its magician master, on Drebbel himself.

The device was relatively simple. It had the shape of a ringed planet with the ring on a vertical axis. The centre was a hollow metal sphere, and the ring surrounding it, a circular tube of glass. The tube was not entirely hollow, having a single closed partition at the top. The sphere was full of air and had an opening that connected to the inside of the glass tube on the left side of the partition. On the right side of the partition the glass tube had an opening to the air. Water was poured into the tube, filling the lower half of the ring. This created a trap that sealed off the left side of the tube connected to the central sphere.

The water in the ring was seen to be in constant motion, first rising up one side and later seen to have shifted the other way. It is nothing more than a fancy version of the inverted retort Drebbel used in his investigations of wind. When the air in the sphere is warmed (passively by the surrounding conditions), it pushes down on water in the ring, shifting its level around to the open side. When the trapped air cools, it withdraws into the sphere and the water swings back the other way.

This was the general description that Drebbel gave the king. However, he had arrayed zodiacal signs and calendar-like markings around the core of the device. If the king had been less impressed with its basic function, Drebbel would probably have lied (as he did elsewhere) that the motion was linked to the ebb and flow of the tides. His eyes were no doubt on the prize – a royal pension – and a tide-predicting machine might well have gone across better if the unveiling had looked to be going south.

So Drebbel's machine basically indicated changes in temperature – what would later be called a thermoscope.

The Dutchman had no interest in quantifying such natural phenomena, but it would soon become all the rage.

❁

Drebbel stayed in England till his death in 1633. After the death of James in 1625, his brand of magic found less favour with the new king, Charles. Drebbel therefore devoted his energies to more practical pursuits in military and domestic technology – seldom with much financial success. In the end he saw out his days as a London publican.

Alongside Drebbel the showman, England was home to another great scientific figure, but one of a very different type. Francis Bacon had served both Elizabeth and James as lawyer and statesman, but he was not at Drebbel's Westminster Abbey display (a large part of why we don't know much about it). Bacon had fallen from grace somewhat, accused of corruption, and was spending his last few years in private study.

While Drebbel's primary goal was to make some cash, Bacon had a larger vision. He planned to recast the total of human knowledge (or what he knew of it) to provide a firm foundation for the society of the future. He called it the Instauratio Magna, which means the Great Instauration, or perhaps better, the restoration. As God had made the world in six days, Bacon was going to correct humanity's inexorable descent into ignorance in six books. It sounds as though Bacon was a bit bumptious, obviously impressed by his own abilities. He died before the Great Instauration was fully complete. Nevertheless he more or less succeeded in his goal.

The second book in the series was the *Nova Organum* ('a new method', in print since 1620). It sets out what would now be understood as the scientific method. Bacon outlines an early glimpse of the process by which observations, hypotheses and experiments can be used to reveal new truths about nature. Within decades, Drebbel's pomp would be long forgotten, and followers of the Baconian method would be casting aside the old-fangled theories of the past and launching the first salvos of the Scientific Revolution.

Bacon's third book, *Natural History* (1622), is a round-up of general knowledge about the natural world. Bacon neglects some of his own advice when it comes to considering the veracity of many statements. For example, he says that a good way of staunching a nose bleed is to dunk your testicles in vinegar. Having said that, perhaps it does work. In other ways he is faultless. He questions whether 'touching Venus' – we can let the reader ponder what that means – really does make you blind. His reasoning is that plenty of eunuchs have sight problems too. But then he goes ahead and says that men 'touch Venus' more in winter, while women do it in summer. This is due to the male form being hot while the female one is cold. So, much of the content bears the legacy of a science that still considers the number of elements to be four. However, *Natural History* does represent a watershed moment in the science of cold.

London and the other great cities of Britain and mainland Europe were at breaking point in the 1600s. Hundreds of thousands of souls were crammed into tiny homes in filthy streets. It was a daily – or nightly – struggle to get enough fresh food from increasingly distant farms into city markets, much of it going bad with no means to preserve it. If a city was to grow and prosper, something needed to change.

Bacon addresses the problem this way:

> *The producing of cold is a thing very worthy the inquisition; both for use and disclosure of causes. For heat and cold are nature's two hands, whereby she chiefly worketh; and heat we have in readiness, in respect of the fire; but for cold we must stay till it cometh, or seek it in deep caves, or high mountains: and when all is done, we cannot obtain it in any great degree: for furnaces of fire are far hotter than a summer's sun; but vaults or hills are not much colder than a winter's frost.*

Natural History contains a list of the causes of cold. For Bacon the *primum frigidum* was earth, not water. Cold can be created by touching another cold body – primarily, one assumes, the ground. Also, materials are intrinsically cold, Bacon contends, especially if they are hard and dense due to a high earth content. The proof was that stones or metals feel cold when touched. Liquids are naturally cold also unless they are mixed with fire, as would be oils and spirits (including the drinkable versions). There are a total of seven causes listed, mostly dealing with the presence of a material substance – although a subtle and unseen one – that makes objects cold.

Bacon also reports experiments into cold, such as putting a sealed bladder (a leather bottle, not an actual organ) in snow or a nitre mixture. The cold makes it shrivel up. In addition feeding a puppy nitre stunts its growth. Bacon says nitre's cold spirit is a hindrance to normal development.

Further, he makes notes on preserving food. He mentions ten methods, including drying and pickling, but his primary means for warding off putrefaction is the use of cold.

It appears that he was still investigating the preservation capacities of cold a few years after the book was completed. In 1626 while travelling out of London, Bacon called for his coach to halt at the base of Highgate Hill, in what was then a farming community north of the city. A snowstorm had hit, and thick snow was settling all around. Bacon climbed to the ground and purchased a chicken from a local woman, paying for it to be plucked and cleaned. He then packed the carcass with fresh snow, in order to monitor it for decay.

However, the 65-year-old scientist was then struck down by the very cold that he was investigating. Too ill to make it home, he went to a friend's mansion in the village of Highgate, where he was installed in a spare room – reportedly in a cold and damp bed that only worsened his illness. Dictating from his bed, too ill to write, he reported that the snow experiment had 'succeeded excellently well'. The frozen chicken was doing better than him and it became clear that he was writing from his deathbed. Ironically the man who began the scientific study of refrigeration was killed by it.

THREE

Applying Pressure

Having six or seven years ago written some tracts in order to the History of Heat and Flame, it seemed the more proper for me to treat of the contrary property, Cold; since, according to the known rule, confronted opposites give themselves a mutual illustration.

Robert Boyle, 1665

For three days and nights in September 1666, London burned. The Great Fire ripped through 13,000 houses, 87 churches and one cathedral, and left almost everyone in the City of London homeless. To the west, the royal palace and Parliament at Westminster had also become alarmingly close to being engulfed. As the wreckage smouldered, the authorities looked to men of science to guide the great city's restoration. As luck would have it, the Bishopsgate headquarters of the Royal Society of London had escaped the conflagration.

The Royal Society had formed three years before, one of the world's first formal bodies devoted to the advancement of knowledge. Its curator, Robert Hooke, was appointed chief surveyor for the rebuilding project. Another academy member, Christopher Wren, collaborated with Hooke to build the Monument to the Great Fire close to where it had started in a bakery on Pudding Lane. True to form, the Monument – an ornate column – was also a scientific instrument, used to measure the extensions and oscillations of weighted springs in its hollow centre and as a fixed telescope for observing the transit of stars. Wren was also given the job of rebuilding St Paul's, the cathedral that came to dominate the city with its enormous dome. Wren used a 26-metre (85-feet) cone made of brick to support the dome as it was being constructed. He reportedly used the bricklaying skills honed in the construction of conical and bottle-shaped icehouses that were becoming common in the grand country estates of Wren's aristocratic friends.

The Royal Society's pool of expertise was also employed in investigating fire itself, not least in the hope of averting similar destruction in future. The year following the fire a newly formulated theory took the lead. It was contained in a book called *Physical Education*, written by a German alchemist named Johan Joachim Becher. Scientific societies like the one in London were still in their infancy, and occultist alchemists were still the leading researchers of the day (many of the Royal Society's early superstars, not least Isaac Newton, were counted in their number).

Becher's *Physical Education* was by no means a guide to health and fitness – alchemists promoted horse manure as a medicine in those days, and a lot worse. Instead the book was a retelling of the nature of solid materials, largely an updating of the *tria principia* of Paracelsus. In place of mercury, salt and sulphur, Becher put *terra fluida*, *terra lapidea* and *terra pinguis*. In so doing he got rid of the supernatural concept of principles and made them physical substances. To balance the books he did away with elemental air and fire. Water remained but the other three elements were three types of earth, or *terra*. *Terra lapidea* was akin to the original hard, stony earth. *Terra fluida* was a volatile spirit, prone to dissipate chaotically (or gaseously) creating flow and motion. *Terra pinguis* was a fatty, oily substance that allowed materials to burn.

Terra fluida and *terra pinguis* were 'subtle', in other words colourless, odourless and impossible to detect directly. However, Becher pointed out that when a substance is burned, the resulting ash is lighter than the original fuel. Here was evidence of the loss of *terra pinguis*, which had been released by the flames. What remained was the ash, *terra lapidea*, which once the final remnants of the other materials had dissipated, became cold, solid and inert.

In the 1650s René Descartes, the idle-bodied and nimble-brained French polymath (he did most of his best work in bed), had proposed a similar theory. Fire was made of minute, perpetually moving particles, which wiggled like eels. The greater their number, the more ferocious the fire and the brighter its light. Coinciding with Becher's theory, cold, said Descartes, was a dearth of the matter of fire. But he also went further. The other physical characteristics of a substance, such as its hardness or fluidity, were due to the motion of its constituent earth and air particles (for him, water was not an element at all).

From its very inception, Becher's theory was found wanting (as was Descartes's). Not every burning substance loses weight. Metalworkers, for instance, knew well that hot iron grew heavier as it was made to glow ever brighter (it was combining with oxygen in the air, but they weren't to know this yet). Nevertheless, the idea would persist for more than a century, growing in acceptance after 1703, when the agent of burning was given the more scientific-sounding name of 'phlogiston' by Becher's student, Georg Stahl.

The theory's earliest and most learned critic was Robert Boyle, another founding figure of the Royal Society, and the only member wealthy enough to pursue research as he pleased. Boyle's work in the 1660s single-handedly transformed the craft of alchemy into the science of chemistry. His 1661 book *The Sceptical Chymist* had debunked the mumbo jumbo of alchemy with a thorough scientific approach. He had already published works on the nature of gas, or 'air', and in 1665 he turned to cold itself, a much neglected subject. He explained in his preface to *New Experiments and Observations Touching Cold*:

> *I remember not, that any of the Classic authors, I am acquainted with, has said any thing of it, that is considerable. They do indeed generally treat of it as one of the four first qualities. But that, which they are wont to say, amounts to little more than that it is a quality*

that does congregate both things of like and unlike nature … Having given us this inconsiderate description of cold, they commonly take leave of the subject, as if it deserved no further handling.

Boyle's handling of his primary subjects – gas and cold – set the clock running for refrigeration technology. After all, right this minute, your kitchen refrigerator is generating cold by the manipulation of a gas. The conquest of cold had begun.

❋

Boyle's 1660 work *New Experiments Physico-Mechanical Touching the Spring of the Air and Their Effects* is the starting point for artificial refrigeration, but this story of 'airs' began 30 years before with a letter sent to Galileo Galilei. The sender was a fellow scientist, Giovanni Battista Baliani in 1630. These two were among the first of what we would now call scientists, although at the time that meant they were part researchers, part problem solvers at the whims of noble patrons.

Baliani's letter related the trouble he was having in building a siphon to draw water over a particularly large hill outside Genoa. At the time it was assumed that siphons worked because they created at least the possibility of a vacuum inside. An actual vacuum was still considered impossible, abhorred by nature in accordance with Aristotle's ancient teachings. However, a potential vacuum was believed to pull on the water and create a flow up and over the siphon. Galileo's reply to Baliani's puzzle was that even the force of a vacuum seemed to have its limits.

Three years later Galileo suffered a meteoric fall from grace. He had pushed the envelope of natural philosophy too far for the Inquisition. Tried as a heretic, he was sentenced to house arrest, where he languished for the remainder of his life. Losing his sight, Galileo employed an assistant, Evangelista Torricelli, to help him finish his life's work.

Among other things, Galileo had been working on a device for measuring changes in temperature. Although not his invention, he would have shown it to Torricelli who later redeployed the idea to address Baliani's siphon trouble.

Galileo's device was not really a thermometer, but a thermoscope. It was less lavishly styled but it worked in the same way as Drebbel's perpetuum mobile. The device was a long tube topped with a sealed glass bulb. The bulb contained air, trapped inside by a column of water in the tube below. If the air was warmed, it expanded and pushed down on the water column. On a cold day the reverse happened; the air shrank in size and the water rose. These fluctuations indicated changes in temperature but were not sophisticated enough to quantify them.

After Galileo's death, Torricelli was given his master's old job as Tuscany's chief scientist. One of the first problems he was asked to solve was the limits of water siphons – no one could build one to raise water more than around 10 metres. Torricelli decided to model the problem, and rather than do it on a full scale, he miniaturised it using a column of mercury. Mercury is a liquid 14 times denser than water. He took a glass tube, sealed at one end like a thermoscope, filled it completely with mercury and placed the open end in a bowl of mercury. The length of the tube was largely irrelevant; the column of mercury inside always sank to the same height (around 76cm or 30 inches in modern units) – a fourteenth of the maximum height of a water siphon.

Although Torricelli died of typhoid just a few years later, the impact of the 'Torricellian tube' was huge. The precise height of the mercury rose and fell from day to day. But unlike in its predecessor, the thermoscope, there was no air bubble trapped inside to push the liquid down. What filled the sealed space above the mercury was something of a mystery. Could it really be nothing at all?

The year after Torricelli's youthful demise, another young genius took over the research. Blaise Pascal had already done enough to rest on his laurels. At 19 he'd invented the first mechanical calculator to help his civil servant father calculate taxes. In later life, following a plea from a gentleman gambler, Pascal collaborated with Pierre de Fermat to develop probability theory, the maths of chance. Pascal even calculated the odds of the existence of heaven and hell and weighed up the risks and rewards of living an ungodly life. He opted for the long odds of a great (and eternal) reward and became the epitome of piety for the rest of his life.

In 1646 Pascal's attention was focused elsewhere, specifically the workings of the Torricellian tube. He had heard about the apparatus from Marin Mersenne, a Parisian friar who was the Enlightenment's version of Google. Mersenne sat at the centre of a web of Europe's scientists, mathematicians and philosophers, and was the chief conduit for scientific communication before the advent of the first scientific journals in the 1660s. Torricelli showed the tube to Mersenne, and Mersenne told Pascal, who had an idea for a now legendary experiment.

Stationed in decidedly flat Paris, Pascal sent instructions to his brother-in-law in the family's ancestral home at Clermont-Ferrand on the edge of the Massif Central mountains. Florin Périer took some persuading, but in 1648 he carried out Pascal's request. One mercury tube was set up at a friary down in the town, and a monk was stationed to watch the level (it didn't move). Meanwhile Périer carried another tube up a nearby volcano, Puy-de-Dôme – fortunately extinct. On the way up, Périer found that every subsequent reading of the mercury was dropping. The higher he went, the force pushing the mercury up the tube was reducing.

Back in Paris, Pascal was delighted by the result. It confirmed his hypothesis that air had a weight. It was the weight of the air – atmospheric pressure – that was defining the height of the mercury. The weight of the mercury in the tube was matched by the push of the air on the liquid in the bowl beneath. As the tube was taken higher, there was less air pushing down from above, so the mercury fell.

Pressure is defined today as a force applied per unit area, and the unit of pressure is named pascal in honour of the young Frenchman. And the Torricellian tube was now firmly established as the first functioning barometer, or pressure gauge. What of the water siphon problem? Galileo's intuition was half right. Siphons have their limit, not from the pulling power of a vacuum but from the pushing power of atmospheric pressure.

Just as importantly, Pascal suggested that the space above the mercury in his tubes was a vacuum. There was simply nothing there. Others disagreed, proposing that Aristotle's quintessence, or aether, must at least be present (though it took Einstein to put this idea firmly and finally into the dustbin of history*). Nevertheless, a couple of years later, in 1650, a German engineer was able to prove that, whatever its precise nature, the vacuum was very real.

Otto von Guericke was a prolific inventor. He produced the sulphur ball, a powerful electrostatic generator that accrued charge through friction (like rubbing a balloon on a sweater). This gadget played an early role in the investigation of electricity. It was also the sensation of fashionable parties for decades thanks to electricians. These were nothing like the same as today's breed; they were showmen who made home visits. Their after-dinner tricks included setting brandies alight with sparks and creating literally electric kisses between lustful guests.

Von Guericke's other invention was more spectacular still. It was an air pump equipped with innovative flap valves that allowed it to extract air from a sealed vessel with great efficiency.

*Einstein's theory of relativity was the final piece in the jigsaw that explained that light and other radiation did not need an ethereal medium to travel through a vacuum, and arrived at all points at the same, constant speed.

Von Guericke came up with many crowd-pleasing ways to show off what the pump could do. The most famous was the Magdeburg Hemispheres, two cast-iron domes that once put together and pumped out showed off the tremendous strength of atmospheric pressure. In a demonstration attended by Emperor Ferdinand III in 1654, two teams of 16 horses harnessed to the hemispheres were unable to pull them apart.

Demonstrably there was nothing inside the hemispheres, and therefore they were being kept together by the same force that drove mercury up the Torricellian barometer. The hemispheres made it clear that nature did not abhor vacuums at all, and when it came to understanding cold, vacuums would be a crucial factor.

❁

Now it was the turn of Robert Boyle. Born in Ireland in the year after Francis Bacon's death, Boyle became an avid Baconian. His early years were dominated by the upheavals of the English civil wars. The Baconian idea of using knowledge to harness the power of nature and further the lot of humanity appealed to the Protestant zeal of the victorious Puritans. However, they were less on board with the attendant concept that proof could only be achieved through experimentation, and anyone who offered a contrary view was treated harshly.

Boyle travelled as a young man and devoted much of his attention to religious writing. This has been interpreted as his way of training his mind, teaching himself to think critically and warding off the lure of a hedonistic lifestyle that would have been easily realised for a young man of his means. Returning home as the first civil war was in full swing, Boyle lay low and devoted his time to scientific research. In these oppressive times, the meetings of men of science were held quietly. Such meetings were good places to share and debate information received from like-minded groups abroad, most notably Marin Mersenne. These gatherings have since been described as the Invisible College, and regarded as a forerunner to the Royal Society, although there is little strong evidence that this was an agreed name used by those who attended. Boyle joined the club in the 1650s.

In 1657 Boyle heard about von Guericke's pump, and by 1659 he had employed a youthful Robert Hooke (one of Britain's most influential and unsung scientists) to retro-engineer one for him. Boyle described his contraption as a 'pneumatical engine' and reported its effects on the 'spring of the air' the following year.

Boyle's wealth allowed him to commission a great variety of glassware apparatus for investigating the effects of manipulating the pressure of air. He proved that air had a weight: a vessel grew lighter as the air was removed. Boyle also found that a feather fell like a stone in a vacuum, that without air a flame could not burn, nor could small animals survive, and that sound did not pass through an evacuated flask. Significantly, a dish of water was seen to freeze as the air was removed.

The Baconian method was meant to uncover unbreakable axioms, laws of nature that underlay phenomena. Boyle's work on air revealed one, now known as Boyle's Law: the pressure of a gas is proportional to its volume. Put simply, pushing air into a smaller space makes its pressure increase. Allowing it to fill a larger space means it will exert a lower pressure.

This is the first 'gas law' of three (although the word gas was not really used for another 130 years). The other two gas laws were not formalised for more than a century. Both involve temperature, which was beyond the scope of science in Boyle's day. Charles's Law (for Frenchman Jacques Charles) says a gas's volume increases as it gets hotter (as long as the pressure remains constant). Gay-Lussac's Law (for Joseph Louis Gay-Lussac, also French) says a gas's pressure increases as it gets hotter (if the volume stays the same).

Boyle was not privy to this fuller understanding of the 'spring of the air', but proposed that air was made of corpuscles, or minute bodies, too small to perceive directly but detectable through their behaviours. The corpuscles were bouncy like springs, and he attributed pressure to their motion. Additionally, he was beginning to think that the heat and cold of a gas were also derived from the behaviour of corpuscles, rather than some kind of *primum frigidum*.

He carried out much of his work on cold in the exceptional winter of 1662, where night-time temperatures were frequently below freezing. However, much of his data was lost when a publishing assistant assigned to transcribe the manuscript ran away to Africa (perhaps for the warmer weather). It took until 1665 for the work to be redone, and it was published in time for the ensuing winter, so readers could carry out their own investigations during the long, cold nights.

Boyle begins *New Experiments and Observations Touching Cold* by making a distinction about using the senses to test theories compared to a scientific instrument. Our understanding of cold comes from our experience of it, but to understand it fully one must note its effects on other objects, especially ones that 'seem more sensible of its changes … less uncertainly affected by them'. Boyle is referring to his sealed weatherglass, a thermoscope about a foot long filled with a column of coloured mixture of alcohol and water contained in a reservoir at the base. Unlike Galileo's air-powered version, hot or cold acted on the liquid reservoir, making its contents expand and contract. Such changes in the 'temper' of the air showed up by a rising and falling of the column. There were no proper gradations to quantify changes, but Boyle was able to take rough measurements, rough enough to show that water that felt cold to the touch was in fact warmer than the air around it.

Deploying his typically impeccable logic, Boyle then set about demolishing the causes of cold proposed through the ages. The *primum frigidum* of Aristotle was water. Manifestly dry materials, such as gold, could also be cold, argued Boyle, and ice forms on the surface of deep water, where it is in contact with air. If water were the primary cause of cold, would ice not form deeper down?

Boyle's inspiration, Francis Bacon, had suggested that earth was the cause of cold. Nevertheless Boyle does not lay into him personally, but picks on the first-century AD scholar Plutarch for his erroneous views: the Earth is coldest at the surface and gets warmer with depth, and becomes more liquid. Surely that indicates that Earth has a hot centre, not a cold one.

Some thought air was the primary cause of cold, including the philosopher Thomas Hobbes, something of a nemesis of Boyle's. Boyle had already witnessed first-hand how air was divorced from causing cold: sucking it out of a flask containing water made the water freeze. Boyle spent a great deal of time refuting Hobbes's claims. Hobbes imagined that cold was carried by the wind. To really drive home his point, Boyle placed living animals in flasks and sucked out all the air. The animals suffocated, obviously, but Boyle's purpose was to isolate them from the wind. He then put the entombed creatures outside, where they froze solid. Hobbes's idea was well and truly debunked.

A fourth cause was also tackled by Boyle. The French philosopher Pierre Gassendi, who had died the decade before, had proposed that nitre contained some physical form of cold, which was why it was so useful in the formation of ice. Boyle dismissed Gassendi's idea because there was no experimental account of his findings. Sea salt and alcoholic spirit have a similar cooling effect as nitre when mixed with snow, so why is nitre the exclusive source of cold?

Gassendi had been pondering the content of the void in a Torricellian tube, and could not accept that it was entirely empty. He proposed that there were forms of matter even in apparently empty spaces. He extended Descartes's idea that heat and fire were the product of fire particles to include particles that spread cold, light and even mediated magnetism and gravity. This concept was also similar to the ideas of the Epicurean school of philosophy dating from the fourth century BC.

Boyle railed against the idea of these hidden corpuscles. He performed experiments to investigate how ice expanded as it froze. This was impossible according to Aristotelian dogma, but mid-winter anecdotes of bottles cracking and water butts bursting said otherwise. First he established basic facts, such as how the volume of the ice was larger than the original water but the weight was the same. Boyle also froze water inside tough metal and earthenware jars filled to the brim and stoppered with a cork. He kept adding weights to the corks to keep them in place, counteracting the force of the ice's expansion. He needed the equivalent of 33kg (72lb) to keep the corks in.

Firstly this proved that cold could not be the flight of Descartes's fire particles. How could something leaving the water, becoming less dominant, produce such a huge expansion force? Secondly, the expansion helped him to explain away the 'swarm of frigorifick atoms' proposed by Gassendi and others ('frigorific' was reputedly coined by Boyle to describe the hypothetical cold matter, and it would be adopted by later researchers in the field). How could such bodies steal into the freezing vessels, passing right through their solid walls without leaving any sign or any damage whatsoever, but once inside create such an outward force that they threatened to crack the vessel apart?

Boyle could not accept any primary source of cold. The only thing that made sense to him was that ice was water that had been expanded by the action of its springy corpuscles, and the process would be the same in all materials (he wasn't to know that water and ice have very particular properties in this regard and that almost all other materials contract when frozen, but we can forgive him that).

Boyle proposed a number of practical applications for cold, at least in terms of scientific investigations. Cold could be used in distillation because while the aqueous portions of a mixture freeze, the other oils or spirits do not. He did not think oils or spirits were capable of freezing. The English winter was never cold enough for him to test this idea. He also bemoaned the imprecision of his weatherglass, seeing the benefit of being able to measure the degree of cold at which freezing occurs for different mixtures and then be able to relate it all to colleagues: 'I consider that we are very much to seek for a standard, of certain measure of cold, as we have settled standards for weight, and magnitude, and time.'

For that, science would need a better instrument: a thermometer.

FOUR

The Temper of the Air

... in winter, air is colder than ice and snow and that now water appears colder than air ... and similar subtleties which the Aristotelian could not explain.

Giovanni Francesco Sagredo, 1615

In 1661 Robert Boyle had got hold of an early form of thermometer. Developed in Florence the decade before, the design is better known today as a Galilean thermometer. It would have been an elegant object, although it was unsuited to measuring cold temperatures and was not that user friendly in the normal temperature range either. Despite the name, this device was not invented by the great Italian scientist himself, rather by his acolytes. This has led to some confusion. Galileo is frequently held up as the inventor of the thermometer. He certainly worked with a thermoscope, but that was also someone else's idea.

In the third century BC, Philo of Byzantium had described a device in which a supply of air was trapped above a quantity of water. Heating the air bubble made it expand, and this change was reflected in a shift in the level of the water. This is the earliest reference to a thermoscope, also known as an air thermometer. Galileo found out about this device via Heron of Alexandria's work, and probably from fellow Italian Giambattista della Porta, who also gives it a mention. In 1593 Galileo is reported to have built a simple version of one using a basin of coloured water and a glass flask.

As the sixteenth century drew to a close, Italy had become a hot market for entertaining trinkets; little gizmos that seemed to harness the spirits of nature. One popular item was a J-shaped tube, which was sealed at one end and filled with enough water to trap some air on the sealed side. The 'spirits' of the air made the water level shift from side to side as the temperature changed. This is all beginning to sound familiar. It is unclear whether the J-tube was inspired by reports of Cornelius Drebbel's perpetuum mobile or vice versa. Drebbel had received a patent for his machine in 1598, and the J-tube craze only really took off across Europe after that.

Whatever the provenance of the idea, Galileo was not very impressed, identifying both for what they were: trickster versions of Philo's original device.

By 1603 Galileo was using an open-air thermometer (similar to the one he would later show Torricelli), which was a glass tube with air trapped at the top by a column of water. The open-ended bottom of the tube was suspended in a bowl of water which was open to the air (the first drawing of one appeared in 1617).

The first person to use the air thermometer for a practical purpose was Santorio Santorii, a doctor from Padua, who was a correspondent and contemporary of both Galileo and della Porta. Santorio, who like Galileo is remembered by his first name, is most famous for proving that the body extracts material from the food it consumes. He did this by making meticulous records of his body weight, the weight of what he ate and drank, and the weight of his faeces and urine. Obviously a man of routine, Santorio did this every day for 30 years.

Santorio also asked the instrument maker (and student of Galileo) Giovanni Francesco Sagredo to make him air thermometers (or thermoscopes) in accordance with Philo's design. It is likely that Sagredo introduced them to Galileo, enthusing in his letters about how they did amazing things, such as show that the air was colder than water, even though it felt the other way around. It is possible that Santorio's thermoscope, which he used to detect fevers in patients, was based on a previous, unrecorded design of Galileo's. However, most commentators give the invention of a practical device that measured temperature to Santorio.

The original thermoscopes had no markings, and no gradations that could be used to quantify changes in temperature. Some reports have Sagredo adding increments to a thermoscope in 1613. He was the first person to refer to 'degrees' of heat and it is thought, therefore, that he divided the tube into 360 units. The number 360 is an easy one to handle mathematically. It can be divided by 2, 3, 4, 5, 6, 8, 9, 10 and 12, and this is why it has been used in measurements since Babylonian times. It's why we count time in units of 60s (a sixth of 360), for example, and why we still divide the circle into 360 degrees.

However, the first documentary proof of a thermoscope with a scale comes from a 1638 document by Robert Fludd, an English occultist. A thermoscope with a scale becomes a true thermometer. Whatever the origin of the scale markings, they were entirely arbitrary, and there was no correlation between devices. For that, thermometer construction needed to be standardised.

The first obstacle was the air. By the 1650s it had become clear that the unsealed air thermometer was sensitive to fluctuations of air pressure, the 'weight of the air' as revealed by the work of Torricelli and Pascal. Therefore it could not be relied upon to measure temperature.

The solution arose from Florence's scientific think-tank, the Accademia del Cimento. This organisation, which was set up in 1657, translates as the 'academy of ordeals', in reference to the stringent experimentation that the members pledged to perform. Their motto was borrowed from Dante's *Inferno*: *'provando e riprovando'*, meaning 'testing and retesting'.

The Academia's patron was Grand Duke Ferdinand II, the great nephew of Francesco I (the lover of iced egg-nog), from whom he had inherited a love of natural philosophy. In true Medici style, ice was a central feature of feasts hosted by

Ferdinand, with drinks served in cups made from ice and tables decorated with ice sculptures.

It is said that Ferdinand himself had made the next breakthrough with thermometer technology a few years before his academy went public. He built a device that was sealed off from the air, and thus immune to changes in pressure. The motion within the thermometer was due to the expansion and contraction of a liquid. For this, Ferdinand settled on using a mixture of water and spirit of wine – distilled alcohol. To make it easier to see the liquid, this was dyed with cochineal, a red pigment made from the crushed bodies of South American bugs (deep red dyes had been beyond the science of Europe and Asia, so this traditional dye, developed by the Aztecs, had become one of Mexico's major industries in the seventeenth century).

Some Florentine thermometers were easily recognisable as such, with a reservoir of spirit beneath a narrow glass column. Boyle's book on cold contains an illustration of this kind of device. However, the accepted design of thermometers from this time was more elaborate.

As the liquor inside Ferdinand's device expanded, it became less dense. By definition, density is a measure of mass per volume. Since the days of Archimedes, it had been understood that the density of a liquid had an impact on what floated in it, and what sank. This principle was put to use via the Florentine thermometer with pleasing effect. Instead of using a narrow tube of liquid, the thermometer was constructed with a wider column. Ingeniously constructed glass bulbs were then floated inside. Each bulb contained a mix of air and oils which gave it a specific density. The oils were probably coloured so each bulb could be differentiated.

When the contents of the thermometer were at a particular temperature (and density), some of these bulbs sank while others floated to the top. The heaviest floating bulb indicated the current temperature. If the conditions became warmer, the liquor expanded and became less dense. As a result, that heaviest floating bulb would sink to the bottom, indicating the rise in temperature. A reduction in temperature would have the opposite effect, making the successively heavier bulbs float upwards as the liquid's density increased.

Ferdinand installed these kinds of devices all around his home and used them to measure the temperatures of bedrooms, kitchens and bathwater. Although the languid motion within these thermometers no doubt fascinated the household, they were hard to calibrate and proved inadequate at low and high temperatures. There was no way of ensuring that the concentrations of water and spirit in each device were always the same, so density changes varied from one thermometer to another. Alcohol boils at a lower temperature than water – about 80°C – around when liquid water has become far too hot to touch. At the other end of the scale, at colder temperatures, the device became clogged up as the water component froze.

The idea of using mercury was suggested. This never boiled or froze as far as anyone was able to tell. However, glassware technology was not able to produce the precision necessary for it. For the next 90 years, the best thermometers were restricted to using alcohol, as pure as possible.

By the 1660s, Robert Boyle and the other Royal Society members were taking their first look at the new technology. Fresh from his service under Boyle, Robert Hooke began carving out his own place in scientific history. His most celebrated work, *Micrographia*, was published in 1665. The main thrust of this was a report of what Hooke had seen through a microscope of his own making. He compared the tiny units that he saw making up the fabric of cork and other plant specimens with the small, simple rooms of learned friars. We have called them 'cells' ever since.

However, the book was not only the foundation stone of cell biology. Hooke also threw in several observations about the hot topics being debated at the time, not least heat itself. Although he was doubtless inspired by his work with 'the incomparable Mr Boyle' (his words), Hooke made one of the first written references to a fundamental concept of heat. In Observation 7 he says that the heat in the substance is from the 'motion or agitation of its parts'. This idea had been hinted at by earlier scholars who were drawn to it by their observations of frictional heating. Plato himself had made the link between heat and motion, and wondered if friction was the origin of all fire. Francis Bacon summed up the heat of friction and the rolling boil of heated water as 'the very essence of heat … is motion, and nothing else'. However, even Hooke's more precise conclusion would be lost again for almost two centuries as those who followed him became intent on pursuing the phlogiston theory of combustion and its resultant heat.

In 1665, Hooke constructed the Royal Society's standard thermometer, the most accurate in England. It was divided into one thousand increments, based on the rate of expansion of spirit, and was used to record weather changes for the next 40 years – in 'degrees Hooke'.

This was not the first formal scale. Santorio had used the heat of a candle flame as his fixed point from which to count temperature changes. The Florentine school quantified temperatures between the coldest frosty nights and the warmest summer days, while Robert Boyle favoured the temperature at which aniseed oil congealed. Edmond Halley proposed the chill of a deep cave as the lower point of a scale with the boiling point of alcoholic spirit as the upper end. However, it was Robert Hooke who was the first person to propose using the freezing point of pure, distilled water – seemingly the most obvious choice.

At the turn of the eighteenth century, Isaac Newton joined the action. Newton was the towering figure of the English Enlightenment. Soon to be the head of the Royal Society, his contributions included no less than the laws of gravity and motion, a telescope design, the colour spectrum and calculus. He defended his claim to scientific advances remorselessly, rubbishing the work of his opponents and leaving reputations in tatters (he'd mellowed a bit by his 80s, which is when the story of that inspirational falling apple came back to him, after a long dinner with his friends and 60 years after it was supposed to have happened).

Legend has it that when Hooke, in his admin role at the Royal Society, had queried some of Newton's ideas, comparing them to his own, Newton's response was his most memorable quote: 'If I have seen further it is by standing on the shoulders of giants.' On its face, Newton's admission appears to be a compliment. Wags have since suggested, however, a more sarcastic reading: being a man of slight build with a debilitating hunchback, Hooke was not one of the giants Newton was referring to. In the coming years the pair would really fall out, which goes some way to explain why the dominant, bullying Newton has remained a famous figure, while Hooke has been sidelined by history.

In 1701, therefore, it would come as no surprise that Newton announced that he had been working on his own temperature scale. Future detective work indicates that he had actually based it on Hooke's standard thermometer. However, Newton makes no mention of this, explaining that he had been working on it all in secret for decades. The Newton scale was meant to work on high temperatures, like the ones found in a chemical laboratory – Newton's research covered alchemy as well as physics and mathematics.

The scale put the 'zeroth' degree of heat at the melting point of snow, and 100 degrees as the heat of a kitchen coal fire. Water boiled at 33°N while body temperature was 12°N. The secretive Newton did not explain how he built a glass thermometer that could withstand the heat of coals. Instead he is thought to have calibrated his scale using his Law of Cooling. This used analytical calculus to extrapolate the original temperature of a lump of iron that had been left to cool for many minutes until it was within the range of a glass thermometer. He had assumed that iron heated to the same red as an ember of coal was the same temperature.*

When Newton went public with his scale it immediately drew criticism from Guillaume Amontons. This Frenchman was a member of the Académie des sciences, the Parisian rival to the Royal Society, filled with several continental researchers who had locked horns with Newton already. Amontons was not looking for a fight, and did not get one because his criticism was so well reasoned. He objected to the use of so many imprecise values for calibrating the scale. Newton had used body heat – 12 degrees warmer than melting ice – to create the degree unit. Body heat was not a rigorous enough figure for Amontons, and neither was the assertion that a summer's day's was five degrees (or perhaps four or six) and that the greatest heat a hand can stand in stirred water was 14 and three-elevenths of a degree. Amontons's review threw Newton's entire enterprise into question.

By contrast, Amontons had made a concrete discovery on the nature of heat and its effects on matter. He placed glass flasks containing different amounts of water and air in a bath of heated water. He found that when he slowly increased the temperature, the water in the flasks all boiled at the same time as the water bath itself. This proved that the boiling point of water was fixed and not affected by the quantity of

*In 1778 Georges-Louis Leclerc, Comte de Buffon, a French natural historian who influenced Charles Darwin and others, used the Law of Cooling to calculate the age of the Earth. He timed how long a ball of iron took to cool from red hot (like the young Earth) to air temperature. He then calculated how long it would take a ball the size of the Earth to cool by the same amount. The answer was 75,000 years, still way off but a marked improvement on 5,782 years, which was the figure promoted by religious leaders.

water being heated. Additionally, and most ingeniously, he measured the pressure inside the flasks as they were heated. They all showed that the pressure of a gas at room temperature had increased by a third after being heated to the boiling temperature of water. Amontons reasoned that gases outside a flask would undergo the same change, thus gas pressure was proportional to gas temperature. This relationship is another of the gas laws. Some call it Amontons's Law, but we mostly attribute it to Joseph Louis Gay-Lussac, who used it to figure out the chemical formula of water (H_2O) in 1802.

Amontons did not stop there. He knew Boyle's Law states that gas volume is inversely proportional to its pressure – the 'spring of the air'. So when he imagined what would happen if his heated gases were able to expand, his conclusion was that they would have increased in volume by a third, because the pressure would be allowed to remain constant. This is in effect the third gas law, which was still 100 years away from formalisation in Amonton's day and attributed by history as Charles's Law. It states that the volume of a gas is proportional to its temperature.

Putting all three ideas together, Amontons formulated a remarkable thought experiment. The simple mathematics of the three laws told him that pressure multiplied by volume was proportional to temperature. Real-world calculations of these variables (that is, figuring out how hot a gas was using values for its pressure and volume) required a constant of proportionality; an unchanging (and unknown) number that did the conversion.

In addition, Amontons asked that if air could expand as it was heated – without limit was the assumption – what happened when it was cooled? Imagining temperatures well below those ever experienced in nature, Amontons pondered the hypothesis that air would condense into a liquid and then freeze. But what if it did not and simply contracted into a smaller volume and exerted a smaller pressure?

Would there ever be a point where the 'spring of the air' became zero? If so, that would result in a zero figure of temperature. Amontons had been deaf since his youth, but his undoubtedly uproarious mind had become the first place that the idea of 'absolute zero' was heard.

In 1702, the same year as Amontons's incredible thought, a Danish astronomer built the first modern thermometer. This was Ole Rømer, already famous in the right circles for making the first empirical measure of the speed of light 26 years before. He had done this with the help of Galileo's four 'Medician Stars' discovered in 1610. Galileo had named them for his patrons, but today we know them as the Galilean moons of Jupiter. Galileo had been the first to spot them orbiting Jupiter – and not Earth (or the Sun) – through his state-of-the-art telescope. The moons' orbits had since been calculated, but they always appeared later than predicted. Rømer realised that the lag was due to the time it took for the light from each moon to reach Earth – and this could be used to measure the speed of light. This was quite a feat. Rømer had no idea how far away Jupiter was, but he used the relative positions of Earth as it spun to come up with a figure. His resulting figure was 26 per cent too slow, but it was a start.

The story goes that Rømer was bored in bed, stuck there with a broken leg. To while away the hours he began to build a thermometer. It is likely that he had been measuring weather temperatures for at least 10 years already, but he was now taking an opportunity to make a better instrument. Rømer spent his time looking for a glass tube that had a constant diameter. He added a drop of mercury at the top and observed its shape as it trickled though the tube. If the droplet thickened or lengthened as it passed along it, that indicated that the tube was inconsistent. It went in the bin, and Rømer moved on to the next. Once he had found the perfect tube he fused a small globe to the end to act as a reservoir – in total it was about 50cm long. He filled it with spirit dyed with saffron and set about calibrating it. His aim was for the spirit to rise by the same length as the diameter of the reservoir for every rise of 10 degrees on his scale.

How Rømer arrived at a zero degree point is not entirely clear. It has been suggested that thermometer makers kept their precise methodology secret to prevent others from providing the same standard of service. He is said to have marked the freezing temperature of water on the tube, and added another to indicate the level of the spirit at the boiling point of water. He then divided the entire tube into eight sections, starting at the upper mark and making the eighth one a fixed distance under the freezing water mark. The upper point was 60, while the freezing point of water was 7.5°Rø. The 0°Rø was meant to represent the freezing point of a mixture of water and salts (as per della Porta), but that might have been a retrospective addition.

Nevertheless Rømer had created a scale that could be applied to other devices, and he made a number of thermometers for recording the temperatures of the air and body. The reputation for such devices brought a young German instrument maker, Daniel Fahrenheit, to Rømer's door in 1708. Retellings of the story sometimes suggest that Fahrenheit tweaked the now elderly Dane's system, but in fact he completely recalibrated it several times over the next 16 years.

Fahrenheit's skill as a glassblower was key, allowing him to make the first working mercury thermometers in 1714. By 1724 he had adjusted Rømer's numbering, removing the tricky half-degree points, and set three fixed points on the scale. The zero degree was the temperature of a mixture of ice, water and a salt. He used both sal ammoniac (ammonium chloride) and sea salt, and suggested that the procedure be done in winter to avoid the ice melting too quickly. This 'frigorific' mixture was the coldest he could consistently produce, although his tinkerings had resulted in much colder mixtures.

The second point was the freezing point of water, set at 32 degrees Fahrenheit, roughly four times that of the Rømer degree. The third point was body heat – most likely the temperature of the mouth – which was 96 degrees. That set the boiling point of water at 212°F.

Fahrenheit's hopes of growing rich from his scale and the devices that used them came to nought. If anything he kept his technology too obscure in an attempt to defend his intellectual property, and as a result his customers did not understand its superiority over rival products. He was to die in poverty, but his name lived on through his temperature scale. Fahrenheit was widely used into the nineteenth century, but by the twentieth it was generally and gradually replaced by the Celsius scale across the world (with the exception of the United States, of course).

The Celsius scale was previously known as centigrade, but had its name changed in 1948. The concept of the scale is more straightforward than Fahrenheit's, pinning the zero to the freezing point of water and 100 to the boiling point. Anders Celsius was not the first person to have this idea, nor even the first Swede; Daniel Ekstöm was among several others who attempted to get the idea off the ground in the 1720s. Frenchman René-Antoine Ferchault de Réaumur used the boiling and freezing points of water, only he divided them into 80 units. He argued that it was simpler than Fahrenheit's scale, more logical, and above all not German. All these attempts nevertheless owed a debt to Fahrenheit's tireless testing and tweaking to perfect mercury thermometry.

Celsius, an astronomer, presented his scale in 1741, having experienced a real-life version of frigid Thule a few years before. In 1736 (the year of Fahrenheit's death) Celsius travelled to Lapland. His reason for going there was not to sample the weather, but to measure the shape of the Earth. A debate of international importance had sprung up between the French and British scientific establishments. Both agreed that the planet was not a perfect sphere. The French said that the Earth was shaped like a lemon, squeezed from the equator so its poles were bulging. The British, armed with the opinion of Newton, said it was more like an orange, bulging out at the equator due to the centrifugal force of its spin. This in turn flattened the polar regions slightly.

The French king sponsored expeditions to Arctic Lapland and Ecuador (named after its location on the equator). Both teams measured the curvature of the Earth and returned (after much hardship) to explain that the French view was incorrect.

While in Lapland, Celsius found it hard to measure the true depths of the cold air and ice. Fahrenheit built thermometers designed for different temperature ranges – the longer ones could take colder readings – but obviously not cold enough. On his return, Celsius created a centigrade scale with the freezing point of water at 100, and the boiling point at 0. His experience of ice had made him more interested in cold than heat, and his scale reflected that.

In 1758 (long after Celsius's early death), the incongruence of this system was fixed. Carl Linnaeus, the *Homo sapiens* botanist who set up the binomial, or two-name, system for classifying life on Earth, flipped the figures so that 0° was cold and 100° hot. If Linnaeus had had his way, we would be measuring temperature in °L: he insisted the scale was his own idea. However, °C, first for centigrade and then for Celsius, eventually became the standard.

❄

As well as recording weather conditions, there was much call for thermometers in the new chemistry departments that were being founded in universities across Europe. Fahrenheit had identified them as a possible market and courted Hermann Boerhaave, one of the leading lights in this field. Boerhaave was a Dutch doctor who lectured in chemistry at the University of Leiden.

Boerhaave's 1732 book *Elementa Chemiae* ('Elements of Chemistry') was filled with information about how the thermometer could be used to investigate the effects of hot and cold on different substances. This caught the eye of Scotland's William Cullen, also a medical doctor (to the king, no less) but, like Boerhaave, one who would be remembered as a research scientist.

In 1746, Cullen became the first academic in Britain to lecture purely on chemistry, when he joined the University of Glasgow. This city would become a key location in the study of temperature, cold and absolute zero, but in 1755 Cullen jumped ship over to Edinburgh, to take up the Chair of Physick and Chymistry.

Much of his research interests lay in cataloguing the melting and boiling points of liquids, not least frigorific mixtures as used by Fahrenheit and 'caloric' ones, which released heat spontaneously. After a year at Edinburgh, Cullen was inspired by a student to investigate the cooling effects of evaporation.

Cullen had tasked the pupil, Dr Matthew Dobson – chemists were generally medical doctors at the time – to investigate temperature changes that occurred when various substances were mixed into liquids. This was an extension of the work of Boyle almost a century before. Dobson (who went on to practise in Liverpool, and discovered the link between sugar and diabetes) reported that when working with the alcohol distilled from wine, his thermometer always dropped a few degrees after being taken out. He suggested that the thin film of volatile liquid on the bulb was evaporating into the air, creating cooling.

Repeatedly applying small amounts of spirit to the bulb made the temperature plunge to below the freezing point of water. It should be noted that it was already decidedly chilly in the Edinburgh lab. These pioneering scientists recorded the air temperature at barely 7°C (44°F).

Further experimentation with a variety of substances – he even tried chilli oil – found that the cooling effect became more marked when more volatile liquids were used. However, Cullen did find exceptions when he used strong acids, which made the thermometer rise. He was quick to conclude that the liquid on the bulb was reacting with moisture in the air, releasing heat that way and masking any cooling from evaporation.

Cullen had also noticed that a thermometer reading dropped when it was left inside a vacuum chamber as the contents were sucked out. After a few moments, the thermometer returned to a reading in accordance with devices outside the apparatus. The drop in temperature was even more marked when the thermometer was coated in alcoholic spirit before the vacuum pump was turned on – greater than a dry thermometer and greater than when a moistened thermometer was left to evaporate in air.

Cullen's hunch was that the vacuum pump was speeding up the rate of evaporation and thus boosting the cooling effect. He decided to explore this further. He filled a vessel with water and placed a smaller one inside it, this time filled with 'nitrous ether'. This was a spirit, a volatile liquid that boiled at a low temperature, made by mixing alcohol with nitre and strong acids. It was used as a medicine at the time, euphemistically renamed 'sweet spirit of nitre' (similar to the stuff Francis Bacon fed to puppies). The pair of vessels was then placed in a vacuum chamber, and some of the air was pumped out until the ether evaporated at great speed. The cooling effect was plain to see – the water in the larger vessel had turned to ice.

Cullen had just invented the first artificial refrigeration system but was unimpressed. He failed to publish his findings in full. However, another of Cullen's pupils, Joseph Black, a rising star of the Scottish Enlightenment, was fascinated with this investigation of hot and cold and the questions that arose. The first one he asked himself was why it was that snow and ice could appear overnight in the streets but often took days to melt away, despite the return of warmer weather. His investigations would find that heat and temperature were not the same thing. In their place would come a new source of hot and cold – caloric – and a new way of understanding matter.

FIVE

Chill and the Airs

I am ignorant of any general occasion or cause of cold, except the absence or diminished action of the sun, or winds blowing from those regions on which his light has the weakest power. I therefore see no reason for considering cold as any thing but a diminution of heat.

Joseph Black, from a posthumous 1807 compendium of lectures

Joseph Black is a Scottish scientific hero, with many discoveries to his name made in that country. And Scotland, specifically its weather, is perhaps one reason for his long contemplations on snow and ice. Black had been born and raised in Bordeaux, where his father was a wine trader. For twelve years he would have enjoyed the pleasant climes of the Garonne Valley before moving to the British Isles to begin his formal schooling in 1740.

By 1754 Black was a qualified doctor, having trained in Glasgow under William Cullen, and taken over the latter's job when he had moved on to Edinburgh. The winters of 1755 and 1756, when Cullen was investigating the cooling effects of evaporation, were much colder than the average, with snow in evidence for many weeks. The average of that time was also lower than today's due to the climate phenomenon now known as the Little Ice Age. This was a cooling of northern Europe (and perhaps other parts of the world) that took place between the fourteenth and nineteenth centuries. But back then it was just the normal weather, and icy conditions were in abundance to be enjoyed, endured and investigated.

Upon hearing of Cullen's first demonstration of the power of refrigeration, Black wrote to ask his favourite teacher to explore the phenomenon of cold further, he himself being absorbed in the properties of air at the time. But after being promoted to become a full professor of medicine at Glasgow University in 1757, he found the time to explore the way water transformed into ice and back again himself.

According to his published lectures, he arrived at this field of inquiry at least in part due to his observations during those long, cold winters.

> *If we attend to the manner in which ice and snow melt … when a thaw succeeds to frost, we can easily perceive, that however cold they might be at the first, they are soon heated up to their melting point, or begin soon at their surface to be changed into water. And if the common opinion had been well founded, if the complete change of them into water required only the further addition of a very small quantity of heat, the mass, though of a considerable size, ought all to be melted in a very few minutes or seconds more, the heat continuing incessantly to be communicated from the air around. Were this really the case, the consequences of it would be dreadful … For, even as things are at present, the melting of great quantities of snow and ice occasions violent torrents and great inundations in the cold countries … But were the ice and snow to melt as suddenly as they must necessarily do, were the former opinion of the action of heat in melting them well founded, the torrents and inundations would be incomparably more irresistible and dreadful. They would tear up and sweep away everything, and that so suddenly that mankind should have great difficulty to escape from their ravages.*

Mercifully for mankind and intriguingly for Black such a rapid liquefaction does not happen. Instead it can take long weeks for thick winter snows to thaw away – and snow remains on high mountains the whole summer long.

Black subscribed to the theory that temperature was the result of a material entering and leaving a substance. He referred to it in his writings as 'caloric', although it was still very much tangled up with the concept of phlogiston, the agent of fire, and treated as more or less the same thing.

Cold, Black was sure, was the product of a lack of caloric. He had no truck with the idea that cold itself was a substance, equal and opposite in its effects to caloric:

> *Many philosophers have thus agreed with the indistinct notion of the vulgar concerning heat, that it is a positive quality, or an active power residing in the hot body, and by which it acts on the cold one; some of them have not been altogether consistent in this opinion. They have not adhered to it, with respect to all the various cases in which bodies of different temperatures act one on the other. They have supposed that, in some cases, the colder body is the active mass, or contains the active matter; and that the warmer body is the passive subject which is acted upon, or into which something is introduced. When a mass of ice, for example, or a lump of very cold iron, is laid on the warm hand, instead of heat being communicated from the warm hand to the ice, or cold iron, they have supposed that there is in the ice, or cold iron, a multitude of minute particles, which they call particles of frost, or frigorific particles, and which have a tendency to pass from the very cold bodies into any others that are less cold; and that many of the effects, or consequences of cold, particularly the freezing of fluids, depend on the action of these frigorific particles. They call them Spiculae, or little darts, imagining that this form will explain the acutely painful sensation, and some other effects of intense cold.*
>
> *This, however, is the groundless work of imagination.*

By 1761 Black had found out why snow remained even when the air around it was considerably warmer. To do so he had first studied the effects of heat on other materials. An early discovery of his was that each substance had a different capacity for heat, in his eyes a capacity to hold caloric.

Black demonstrated variations in 'heat capacity' by chilling a lump of iron and a block of wood in the plentiful supply of snow afforded by life in Glasgow at the time. After leaving them for a while to cool down, he then picked them up and held one in each hand. The metal felt colder than the wood, and as we have seen, Black correctly reasoned (partly) that it was drawing the heat, or caloric, from his hand more readily than the wood. It also meant that the metal had a low heat capacity (known better today as specific heat). In other words it only needed a little added caloric to make it rise in temperature. Water by contrast had a much larger heat capacity. Heat seemed to act slowly on it – the water could take in a lot of caloric without significant rise in temperature.

Black's next phase of research was dependent on the accuracy of thermometers, which was not necessarily proven at that point. Therefore, he segued his investigations into how materials were altered by heat. He showed that almost everything that could be heated (and not burned) – mostly metals and liquids – expanded when hot and contracted while cooling. Crucially, though, their weight stayed the same (Black had already made his name by inventing an ultra-precise design for weighing scales when he was just 22). Brittle objects tended to crack, he observed, because heat was not entering them consistently. While warmer parts expanded, others did not, and so they broke apart.

The exception to this expansion rule was, of course, ice, which expanded when it froze. Black's views on this would have relied on the research of Edme Mariotte and Jean-Jacques Mairan. Mariotte had been a contemporary of Robert Boyle, and Mariotte's Law is a Francophone version of Boyle's (it says exactly the same thing, but is given a French attribution). Mairan's other work on evaporation, meanwhile, had been one of the inspirations behind William Cullen's prototype refrigerator. Both these Frenchmen had measured the expansion of ice, something many of their contemporaries had also considered.* Mairan had found

*Christiaan Huygens, a Dutch scientist, had even cracked open the mightiest cannon barrel using the expansive force of ice.

that when water that had been boiled froze, its volume increased by just a few hundreths. The boiling had driven out any air dissolved in the water. However, ice made from unboiled water – still filled with air – expanded by almost a tenth (the difference is clear: pure water ice is an eerie blue, while most of the stuff we see is clouded white with all the minute air bubbles in it).

Black reasoned that the expansion was in part due to the presence of air, but also hypothesised that it was due to some unknown mechanism by which the particles of water were being repositioned into a bulkier arrangement (this is indeed what is happening through the action of intermolecular forces that would not be fully understood until well into the twentieth century). In later lectures on the subject, Black included the discovery of Frenchman Antoine Baumé, who had found that water contracted as it cooled (like anything else) until it was near to the freezing point (the modern figure is about 4°C; Baumé's was a bit out), then expansion began even while the water was still in its liquid form.

The research into expansion and contraction helped to convince Black that the movement of mercury in his thermometers did accurately reflect changes in temperature. He was now ready to record the way water and ice changed their temperatures. This involved testing a concept now known as thermal equilibrium. Black was first to set out the basic idea:

> *Even without the help of thermometers we can perceive a tendency of heat to diffuse itself from any hotter body to the cooler around, until it be distributed among them, in such a manner that none of them are disposed to take any more heat from the rest. The heat is thus brought into a state of equilibrium.*

An experiment by the English mathematician Brook Taylor in 1723 had already confirmed this, and was repeated by Black in 1760. One pound of hot water mixed with one pound of cold water resulted in two pounds of water with a temperature halfway between those of the original constituents. It sounds so obvious today, but without the painstaking development and exhaustive proofing of thermometers no one was sure of it until Black's work.

Obvious it may be, but when ice is involved it nevertheless becomes somewhat baffling and here's why. When ice melts – on a mountaintop, in the street or in the chilly labs of Glasgow University – it is entering a thermal equilibrium with the air around it. Compared to the sample of ice there is a much larger quantity of air, so quantifying the fall in air temperature to match the rise in ice-water temperature is less than easy – and was impossible in the 1760s.

Instead, Black compared the time it took for ice and water to reach equilibrium with the air. He took two identical flasks and filled both with identical amounts of water. One was frozen using a frigorific brine and the other was cooled to as close to the freezing point as possible. The assumption was that the temperature of the contents of the two flasks was as close a match as could be achieved. Both vessels were then quickly carried to a large hall kept free from draughts and any other disturbance. They were both hung in the centre of the room, close together so that they were exposed to the same temperature of air – but not so close that they would affect one another. Then they were left to warm up.

Black says that the hall was kept warm but nevertheless the air was 47°F, which is a less-than-toasty 8°C. In half an hour, Black reports, the water flask had reached 40°F (around 4°C). At this time the ice flask was mostly water, although there was a bit of ice left to melt. Black took the temperature of the water nearest to the edge, away from the lump of ice persisting deeper in. It was also 40°F. However, Black suspected that the heat of the air was not getting through the water to the ice very quickly. He was right. It took ten and a half hours for all the ice in the flask to melt and for the water to reach a temperature of 40°F.

Although the rises in temperature had been roughly the same, the ice had needed an influx of 21 times more heat to do it. Black's counter-intuitive discovery was that heat and temperature were not the same.

Next, Black came up with a quicker experiment, although it was a good deal more fiddly. He took a sample of ice and waited until it was wet all over, showing him that it had reached melting point. Then he scooped it up in a woollen glove to insulate it from the heat of his hand and weighed it. Actually, he hastily balanced it on the scales with an equal weight of sand and returned later to measure the sand's mass. Then the melting ice was dropped into a quantity of water – again precisely measured – which had been heated to 190°F (88°C). The ice melted within seconds and the resulting water temperature was taken.

The difference in temperature between the two components was 158°F, but Black had to account for the heating of the glassware and the fact that the ice was not the same mass as the water – although not too far shy. His calculations suggested that if all the heat in the mixture was to be observed in a temperature rise, it would go up by 86° on the Fahrenheit scale. It went up by just 21°.

Black now made his big step. He described the heat that can be measured as temperature – or felt on the skin – as 'sensible' heat. However, some heat was hidden from the thermometer by the ice as it melted into water. This he called latent heat.

To triple check, Black then ran the process in reverse, making use of the sharp winter frost. Two identical volumes of water were left to chill in the cold air. One was pure water, the other was mixed with salt and alcohol to lower its freezing point, an Enlightenment version of antifreeze. The mercury thermometers in both plummeted in unison until the water hit its freezing temperature. Then the thermometer stayed at that level while the ice formed around it. Meanwhile the briny spirit continued to cool down at the same unchanging rate.

Black's theory was that heat – the caloric matter – was being lost from both at the same rate, but the heat being lost by the water as it was making the transition to ice was latent. Black imagined that caloric surrounded the particles (he used the word 'atoms' for this – though it should be noted that atoms were purely theoretical entities in those days). The pause in cooling captured by the thermometer, Black explained, was because liquid water 'atoms' held on to a much larger cluster of caloric particles than when they were in the form of ice. In order to freeze, the clusters had to be thinned out considerably. Only then did the loss of further caloric result in a drop in sensible heat.

To melt, the ice particles had to gather up enough caloric to become water. And Black showed that the same process was happening when water boiled and condensed. To go from water to steam required an even greater influx of hidden caloric.

Although the idea of frigorific matter was still being proposed by some at the time as a separate substance that caused cold, the caloric theory (with its cousin, phlogiston) was the only show on the road entertained by most scientists. The question was what kind of stuff was it and where did it come from? To answer that, a whole new cast of characters would take to the stage.

❅

The next act has an easy opening scene with a familiar face. A new field of science had begun, in which air was finally shown to be a mixture of gases and not an element at all. This new family of substances would crack the problem of combustion and explain why things burned, and with that forward the study of heat. The science was called pneumatic chemistry, and its founding figure was none other than our friend Joseph Black.

While still a medical student under William Cullen, Black chose to investigate kidney stones for his doctoral thesis. Specifically, he wanted to find a medicine that could remove them without the need for an operation. Surgery was an accepted part of medical training by this time, although it was still regarded by some as a manual skill rather than a knowledge-based one. The Royal College of Surgeons of Edinburgh had only recently split from the College of Barbers, after all. Practical anaesthetics were many years away, so something as common as a kidney stone would often require an excruciating procedure, during which the patient was conscious throughout – and then left to fight off any infection resulting from the incision.

The undergraduate Black's idea was to use the new science of chemistry to remove the stones. He began to search for a medicine that could dissolve the stones in situ, and his first candidate was quicklime. Black knew that quicklime could destroy a kidney stone, but he also knew that it destroyed soft tissue as well. We now understand quicklime as being calcium oxide and this reacts with water, even taking that from the air, to form calcium hydroxide, or slaked lime. That process releases a lot of heat. When quicklime comes into contact with body tissue, the reaction makes use of any moisture in the tissue and results in serious burns.

This fact alone would have turned Black against pursuing quicklime as his kidney-stone cure-all, but it is also worth noting that in those days no one really knew what quicklime was. Black's supervisors could not agree on how it should be made. One said it should be made from the 'ash' of oyster shells, while another insisted it be made by 'burning' limestone. In fact both processes produce the same stuff, but back in the day when chymists were chymists, not chemists, no one was to know that an identical substance could be produced using seemingly different raw materials.

Instead, the wise young Black opted to side-step the problem by taking magnesia alba as his subject. The term alba indicated that this was a white crystalline substance related to magnesia, a mineral or 'earth' well known since ancient times. It was named after a region of northern Greece, where once was an ancient kingdom of that name. We also get the word magnet from that root because the area is rich in lodestones.

Preparations of magnesia had been used for medicinal purposes for centuries, most notably as a laxative. Black takes note of an earlier researcher's work on magnesia in his records: 'Magnesia appears to be a very innocent medicine; yet, having observed that some hypochondriacs, who used it frequently, were subject to flatulencies and spasms, he seems to have suspected it of some noxious quality.'

Magnesia is a rock-forming mineral, a natural form of magnesium oxide that is common in marbles. When swallowed it reacts with stomach acid, settling any indigestion, and also acts to draw water into the colon which helps to loosen everything up. A milder form developed in the nineteenth century, containing magnesium hydroxide, does the same thing. We know it as milk of magnesia.

Magnesia alba appeared to be drawn from the same family of substances, although it had a distinct character; the field was far from precise. The name for magnesium metal obviously derives from its presence in these minerals, but magnesia negra, a black earth hailing from the same area, contains manganese, an entirely different metal (albeit one with the same name-derivation).

Black had heard that magnesia alba could be made using the left-overs from nitre production, and also from the salts remaining after seawater was boiled away. However, he obtained his samples through a convoluted extraction process from Epsom salts, another crystalline folk medicine that is named after a large deposit in the hills to the south of London.

Once purified, Black found magnesia alba to be an alkaline earth, although a mild one. In those days, every crystalline powder was still described as an earth (similar solids that resisted pulverisation were known as stones). The term alkaline referred to its action when mixed with an acid. Alkali is derived from the Arabic for 'the calcined ashes', which goes back to the early days of alchemy when active chemicals were made from wood ash.

Despite its ancient provenance, alkali is still used today to denote a soluble chemical that will react with an acid to create a salt. Putting aside our use of the word salt to mean the stuff we sprinkle into food, chemists understand it as an inactive and neutral material, which is still in keeping with the definition of Paracelsus even to this day.

Black soon found that magnesia alba was of no use in treating kidney stones, but one suspects he did not mind that much. His fascination with chemistry got the better of his need to extend a medical thesis. In fact his lack of attention to his medical studies meant he found it hard to find work on graduation, and instead had to rely on teaching to make ends meet – something he proved to be very good at.

Black found that adding acid to the magnesia alba crystals made them bubble, giving off 'air'. The acid he used was spirit of vitriol, an old name – and a much better one – for sulphuric acid. Next, he heated a sample of magnesia alba. This had no discernible effect – the powder was still white – but the addition of acid no longer produced the bubbles. Black's hypothesis was that heating had removed a component of the magnesia alba that gave rise to the production of 'air'.

The next step was to collect the 'air' being released. This proved impossible because the standard method was to funnel it through a water flask so that it collected as a single large bubble at the top. The mysterious air simply disappeared, dissolving in the water.

Black decided on a new approach. He weighed a sample of magnesia alba before it was heated, then again afterwards. It had gone down in weight, and he reasoned that the loss was due to the release of the unknown air. He did the same with a mixture of acid and magnesia alba, noting their combined weights before and after the reaction had ceased to release the gas. The two experiments yielded the same results. Therefore the action of acid was having the same effect as the heating.

Black then ran the whole thing in reverse. He heated some magnesia alba to drive off the air and dissolved the residue in acid. Then he added another mild alkali – probably chalk or dried limestone – to the mixture. The second alkali caused the same kind of fizzing, and when the resulting mixture was analysed, Black found that it contained magnesia alba again.

Black now gave a name to the gas in the bubbles – fixed air. With this term he proposed that some component of the air was absorbed or 'fixed' by the magnesia alba, only to be released again by heating or a reaction with acid. The same 'fixed air' was being released by other mild alkalis, and Black also found it present elsewhere.

While the gas had no effect on pure water it did do something to limewater. Limewater is made by adding a large amount of water to quicklime so that the solid dissolves completely, creating a liquid that looks more or less like pure water. When the fixed air was bubbled through limewater, it made it go cloudy. Smoke from a wood fire was already known to do the same thing. Black saw this as evidence that his fixed air and the action of fire were linked in some way.

The modern explanation of all this is straightforward enough. Magnesia alba is magnesium carbonate. Heating it makes it decay into magnesium oxide and release carbon dioxide gas. When magnesium carbonate reacts with an acid, in this case sulphuric acid, it forms a salt, magnesium sulphate, some water and some carbon dioxide gas. This gas – Black's 'fixed air' – seen bubbling out of the reacting mixture is exactly the same stuff as is driven off by the action of heat.

All carbonates, known to Black as mild alkalis, will do the same thing. Limestone is mostly calcium carbonate, as are the shells of oysters and other shellfish. Heating the calcium carbonate makes calcium oxide, or quicklime. Adding that to water makes slaked lime, calcium hydroxide, and this is the active component of limewater. Calcium hydroxide combines with carbon dioxide to make calcium carbonate again, and tiny specks of this solid form in the water, giving it the tell-tale cloudy, milky appearance.

Compounding his discovery with yet more, Black found that the air, or gas, produced by fermentation also contained fixed air, as did the breath we exhaled. To test this further he placed birds and mice in jars of fixed air, and watched as they died of suffocation. Unlike ordinary air, fixed air could not support life.

The meticulous experiments had shown that the changes in weight of the solid samples were all due to the loss and addition of fixed air, nothing more. Nevertheless, Black added phlogiston to the mix when he came to explain the process.

Remember, the phlogiston theory of fire had been proposed 90 years before based on the observation that weight was lost during combustion – this was phlogiston being released. Later on, it was found that some substances, such as metals, gained weight when burned. It takes a great deal to incinerate a metal, so it was proposed that a great deal of phlogiston was given out in the process, so much so that the weight of the products went up! The implication was that phlogiston had some kind of property of anti-weight, or levity.

On the face of it, it is perhaps surprising that someone as rigorous as Black would refer to a theory that was so manifestly on shaky ground. However, Black had no other word to use to describe the action of heat in his experiments, so phlogiston made its way in. Nevertheless, Black's understanding of phlogiston was more nuanced than simply that it was the stuff of fire that burst out as flames. Instead he suggested it could act in other ways, altering the nature of substances without necessarily setting them on fire.

Black proposed that phlogiston entered a mild alkali when it was heated, and that forced out the fixed air. The same process of heat acting on a burning substance was also releasing fixed air – and this was the cause of the loss of weight. To Black phlogiston was not merely an agent of fire that produced flames. He also used it as an agent that could make other kinds of change to substances, something akin to what might today be simply referred to as 'energy'.

Joseph Black was quick to drop the phlogiston theory when an alternative was proposed, but that would have to wait until several more 'airs' had been added to his very first discovery.

❋

Within a decade of Joseph Black's publications on 'fixed air' in 1756, a young scientist from London had found another 'air' which also seemed to be locked away inside solids. The researcher was Henry Cavendish, a member of an aristocratic scientific family. The young Henry was socially awkward, to say the least, and much of his early career was sponsored by his father, Lord Charles Cavendish, who secured him a membership of the Royal Society at the age of just 21. Son Henry lived and worked at his father's London home, building a laboratory that was connected to his living quarters by a private staircase. He spent a great deal of time alone, communicating with family and servants alike through notes. Friends did not come until much later, if at all, with Henry being one of the most regular, but also the quietest, attendees at Royal Society functions.

The Cavendish scientific connections were not put to waste, however. As an older man, Henry Cavendish succeeded in 'weighing' the Earth with a device that measured the gravitational constant used in Newton's universal law. However, his contribution to our story began in 1766 when he revealed his initial work on 'air' released from metals.

Just as Black had done, Cavendish began his research by observing how iron created a gentle fizz of bubbles when dropped in strong acid (in fact many other metals do this as well, but in the eighteenth century only inactive metals, such as copper, silver and tin, were readily available, and they don't). Unlike the 'fixed air' of Black, the gas observed by Cavendish did not mix well with water, so he was able to collect it in large quantities by directing it through a cooling water bath and letting it collect in glass flasks or paper balloons.

He called the gas he collected 'inflammable air', because when lit with a flame a sample of gas burned with great speed, a yellow flash of flame and a loud pop, or squeak. 'Inflammable air' was also found to be very light indeed, almost weightless when compared to the ambient air.

Was this fast-burning, super-lightweight material pure phlogiston, the very agent of fire? With that question left hanging, 'inflammable air' withdraws behind the scenes for a while, like the reclusive Cavendish, and two more 'airs' take centre stage.

*

A lasting legacy of Joseph Black's work on 'airs' was that fixed air came to be seen as some kind of 'bad air', a foul by-product of fire, including the fires that kept life burning in the body. Therefore it was assumed that there was a 'good' component of air that supported burning.

In 1772 science history repeated itself. Daniel Rutherford was a medical student training under Joseph Black, who by this time was teaching at Edinburgh University. As Black had done before him, Rutherford chose to study air for his thesis.*

Rutherford had shown that a candle flame died when confined to a sealed flask of air, and that a portion of the original air inside had been converted to fixed air. From this he developed a hypothesis that 'bad air' contained some kind of factor that would prevent combustion or life. His suspicion was that fixed air contaminated the rest of the air, making it 'bad'.

To test this he removed the 'good' air from a sample by burning a candle in an upturned flask as before. Next he bubbled the air left behind through some limewater to extract the fixed air. However, the air that remained after all this was just as 'bad' as before. It snuffed out a flame and killed any creature unfortunate enough to be put inside the flask.

Although Rutherford did not realise it, he had isolated nitrogen, which makes up the bulk, 78 per cent, of the Earth's atmosphere. Rutherford did not recognise it as a new chemical substance. Instead he deployed the phlogiston theory to make sense of it. His name for the gas was 'phlogisticated air', and the idea behind it was this – burning the candle released phlogiston into the air that was trapped inside the flask. At some point the air became saturated with phlogiston and could absorb no more; it was phlogisticated. Thus combustion ceased.

The next gas to be discovered was the 'good' part of air, and it would be treated in much the same way at first. The prime mover in this discovery was Englishman Joseph Priestley, a preacher turned chemist who had recently won the Royal Society's top award for a world-changing invention. Priestley's achievement was not a new engine or powerful instrument. It was soda water.

*It is perhaps also telling that he, like his mentor, went on to become an academic *in lieu* of a medical career, eventually specialising in botany.

After travelling to Leeds in search of a congregation in 1770 (he did not find one), Priestley had taken lodgings next door to a brewery. As Black had mentioned fermentation as a source of fixed air, Priestley used the brewery as a makeshift laboratory to indulge a new-found hobby of chemistry. It is reported that he confirmed that the gases rising from the fermenting brews did indeed contain fixed air by dangling animals over the vats. A butterfly, frog and mouse were all overcome by the fumes, providing Priestley with his proof. Like Black he also found that the gas dissolved readily in water, but he then went further and drank some of the mixture. The result was a surprisingly refreshing drink. As well as the top science prize of the age, Priestley's 'soda water' awarded him access to English high society (it took Johann Schweppe, a Swiss jeweller, to turn soda water into the multibillion dollar global industry that it is today).

Priestley was invited to be the secretary and intellectual companion of the Earl of Shelburne. Installed in the earl's country home at Bowood House in Wiltshire (where his little laboratory remains on show), Priestley had much more time to indulge in pneumatic chemistry. His discoveries were later recorded in his work *Experiments and Observations on Different Kinds of Air*.

The first thing Priestley did was develop a means of collecting fixed air by bubbling it through mercury instead of water. He then used the same technique to collect nitrous air. This gas is given off when a metal, often copper, is dropped in nitric acid, or spirit of nitre as Priestley knew it. Today we understand this colourless gas as nitric oxide.

Priestley then found that candles burned more brightly when placed in nitrous air that had been treated with iron filings and brimstone (the old name for sulphur). He called the resulting substance 'dephlogisticated' nitrous air. In the context of the phlogiston theory, he believed that the original nitrous air was near saturated with phlogiston and thus did not support burning. However, the treated air had given away its phlogiston to the iron filings. Thus when a candle was placed inside, the dephlogisticated gas around it had more capacity to receive phlogiston from its flame.

What was really happening was that the iron was converting the nitric oxide gas, which is made from one nitrogen atom and one oxygen atom, into nitrous oxide, which has two nitrogens for every oxygen. That was releasing pure oxygen into the mixture, and it was this that was making the candle burn so brightly. Priestley didn't know it yet but he had made pure oxygen.

Chemistry still had some way to go to reveal the role of oxygen in burning. The next clue came from mixing nitrous air (nitric oxide) with air. It spontaneously decomposed into a haze of red-brown fumes (this is nitrogen dioxide, the stuff that makes car exhaust fumes brown). The reaction occurred with no need for flames or other heat, and Priestley noticed that once completed the volume of the gas was always reduced by a fifth. This reduced volume of gas was also phlogisticated – in other words, nothing burned inside it.

Priestley grasped that the missing 20 per cent of volume must be the 'good' component, or the 'dephlogisticated' air. However, it took him almost two years to make a sample of this dephlogisticated air, and he stumbled upon it more or less by accident. In August 1774, Priestley produced the gas by heating mercuric oxide, an alarmingly bright orange and exceedingly toxic chemical. Just like the dephlogisticated nitrous air, the novel gas was shown to boost the flame of a candle and even make hot charcoals flicker back into glowing embers. Even when mixed with nitrous air, creating a puff of red vapour, the remaining gas still supported burning. That sealed the deal – this gas was pure, dephlogisticated air.

Priestley confesses to his surprise at the result:

I cannot, at this distance of time, recollect what it was that I had in view in making this experiment; but I know I had no expectation of the real issue of it. If, however, I had not happened for some other purpose, to have had a lighted candle before me, I should probably never have made the trial, and the whole train of my future experiments relating to this kind of air might have been prevented.

History tells us that Carle Scheele, a Swedish apothecary, had isolated a gas that he called 'fire air' in 1772. This is now confirmed to have been oxygen as well. However, Scheele had failed to communicate his discovery to the wider world until it was too late. When news of Priestley's find spread, those who heard about it believed the Englishman when he said, 'Who can tell but that, in time, this pure air may become a fashionable article in luxury. Hitherto only two mice and myself have had the privilege of breathing it.'

*

Within three months of his discovery, Priestley found himself en route to France with his patron, Lord Shelburne. They were heading for an audience with Antoine Lavoisier, one of the most prolific and certainly the wealthiest chemist at work at the time.

Lavoisier's wealth allowed him to amass a great collection of bespoke apparatus for his investigations. He had entered the fray in 1770, while still in his early twenties, and made his mark in the Acadamie de sciences through his use of precise measurement – thanks in part to his elaborate equipment.

One of his early successes was to disprove a theory, widely held at the time, that water was gradually transmuted into earth by boiling. The inspiration for this theory was that a thin residue of minerals built up when a quantity of water was repeatedly distilled. The question was, did this solid come from the water itself or from an external source? Lavoisier used a sealed glass vessel called a 'pelican' to find out. This had a large base chamber connected to a bulb-like upper chamber by a slender neck. The upper and lower chambers were then connected by hollow handles. It did not look anything like a pelican, but the principle was that vapours rising from a boiling mixture in the lower chamber would condense in the upper one, and the liquids would trickle back down the handles to the lower section again. Lavoisier boiled three pounds of water in the pelican for 100 days. A residue did indeed form on the bottom, and when he weighed everything inside – and the pelican – he found that the weight of water was unchanged, but the weight of the residue matched the weight loss of the pelican. A bit of precision was all that was needed to show that water did not transmute magically into earth. Instead the powerful heat from boiling water over 100 days had dissolved away some of the glass.

In 1772, Lavoisier began to investigate the process of combustion. He had no truck with the phlogiston theory, and repeatedly showed that when substances burned, part of the air was removed – all recorded with his usual precision. The meeting with Priestley in 1774, in which the Englishman outlined the properties of his newly discovered gas, would prove a turning point.

Lavoisier was a great one for stealing other people's ideas, or at least relating their discoveries without proper attributions, thus giving the impression that they were all his own work. Having said this, as with the airs of Rutherford and Cavendish before him, Priestley did not regard his 'dephlogisticated air' as a separate species of chemical (in fact he was still formulating his dephlogisticated concept). Instead he regarded it as a very pure kind of air. It was Lavoisier who saw it as a distinct substance, indeed a new element, along with other gases (it is thanks to Lavoisier that we can finally give up the term 'air' and use 'gaz', or 'gas', instead).

In April 1775 Lavoisier presented an early form of his combustion theory, which related how heated metals were combining with a portion of the air. When these 'calcined' metals – we would now understand them as oxides – were smelted or 'reduced' to pure metals in a charcoal furnace, they released the air again, this time as fixed air. He reported that if the reduction of metal was achieved without the use of charcoal – presumably he meant reducing mercuric oxide using the same method as Priestley – then the air released was the same pure material with which the metal had combined in the first place.

Lavoisier's report on this subject is known as the *Easter Memoir*, and he made several changes to it over the next three years. So by 1778 a revisionist version of his 1775 'discoveries' made a distinction between atmospheric air and the gas involved in combustion. By 1779 he had a name for it, oxygen, which means 'acid-former'. The reason for this name is based on a misconception by Lavoisier; oxygen has no role in acids, but nevertheless the name stuck.

Oxygen makes up a fifth of atmospheric air, while the remainder is phlogisticated air, which Lavoisier wanted to call azote, meaning 'lifeless'. The name nitrogen – 'nitre former' – was settled on in the 1790s for this relatively inert gas (and a century later, nitrogen was one of the first gases to be chilled into a liquid; its liquid form has been an important mediator of deep cold ever since).

By 1783, Lavoisier was able to disprove the phlogiston theory once and for all. Combustion was simply a reaction between oxygen in the air and another substance. For the avoidance of doubt, Lavoisier was able to show that even water was the product of combustion. For this he made use of Henry Cavendish's inflammable air. This was placed in a flask with oxygen, and the two gases were sealed with mercury before being ignited with an electric spark. The flash of combustion produced a vapour that rapidly condensed on the inside of the glass and formed a pool of liquid water floating on the surface of the mercury. Inflammable air was renamed hydrogen, 'the water-former'.

So neither air nor water were indestructible elements, and earth had long been understood as a mixture of many distinct substances. And now fire was explained away as a chemical reaction. Where did that leave heat?

Lavoisier encapsulated his answer in a list of 33 '*substances simples*', materials that could not be broken down further into simpler constituents. We know them today as the chemical elements. The list, one of the first of its kind, included gold, iron, sulphur and carbon, and of course the three new gases, hydrogen, azote (or nitrogen) and oxygen. However, at the very top were *lumière* and *calorique* – light and heat. To Lavoisier these were substances as well. The next thing he was going to do was measure them.

SIX

Going for the Motion

Three feet of ice is not due to three days of cold.

Chinese proverb

In order to demonstrate the transmutation of gases into water, Antoine Lavoisier had recruited the help of his friend, Pierre-Simon de Laplace. A fellow aristocrat, Laplace was at least the equal of Lavoisier's intellect and a good deal more level-headed. He made contributions to the philosophy of free will, mathematics and astronomy. Concerning the latter, one of his notions was of a body so small and dense that its escape velocity – the speed required to break free of its gravity – would be faster than the speed of light. This '*corp obscur*' was one of the first formulations of what we now think of as black holes.

Laplace was also a successful politician, managing to steer his way through the court of Louis XVI, survive the tyranny after the 1789 French Revolution and then serve as a minister in Napoleon's government. His colleague Lavoisier was less adept in such things. His situation after the revolution was difficult to say the least, because his wealth was accrued through 'tax farming'. He had lent money to the king each year and was repaid with a profit by collecting taxes from the imports of commodities such as tobacco and salt. Although consumed by his science, he did voice concern about the inequality, albeit in a very aristocratic way. A happy peasantry, he counselled, was the best way to boost productivity and the taxes that came from it. There is no doubt that he had little idea of what life was really like for his compatriots and had no intention of finding out.

This attitude was ably illustrated by one of his public displays of combustion at work. In a celebration of the new science and of his great ingenuity – and great wealth – Lavoisier built a solar furnace. This was a giant version of the heating glass chemists used in those days to warm concoctions with a focused beam of sunlight. Its purpose was to provide a source of great heat, and to do it without the need for burning a fuel, which might otherwise introduce contamination. Lavoisier's furnace was the size of a bus, with a main lens 132cm (52 in) across made from curved sheets of glass filled with spirit vinegar. It was said to be able to melt platinum on a sunny day.

In 1772 Lavoisier had shown off his furnace with an experiment that in hindsight was the epitome of the problems facing France at the time – although also a fascinating bit of science. He procured a diamond, a stone that was even more expensive in relative terms than it would be today; it was so expensive that he had to club together with a few other wealthy gentlemen to buy it. He then simply incinerated it.

The pricey gem disappeared and in its place was a vessel of fixed air, or carbon dioxide. Lavoisier had provided proof that diamond was a crystalline form of carbon, but at what cost? In 1794, at the height of the Reign of Terror that followed the revolution, Lavoisier's aristocratic past caught up with him, and the guillotine took off his brainy head.

Although cut down at just 50, Lavoisier had been able to make one further step forwards in the science of heat and cold in the decade leading up to his death. In partnership with Laplace, he perfected a device he named the calorimeter. It was the first of its kind, designed to measure the amount of heat given out by chemical reactions, and it made use of Black's concept of latent heat and his own theory of combustion. Modern devices with a similar purpose but different designs are still called calorimeters today.

The Lavoisier–Laplace instrument was a marvellous piece of engineering. On the outside it looked like a small metal vat, standing on a tripod with a conical bottom. Inside, sealed under a heavy iron lid, were two more concentric chambers. The inner one was the reaction vessel. It was a wire-mesh basket designed to accommodate a number of different bits of apparatus, such as a gas flask or crucible. It was even used to hold animals. The central chamber was suspended in a middle compartment packed with snow or pulverised ice so it completely surrounded whatever was being analysed. The base of this ice chamber had a small grill through which meltwater could flow. This was tapped off from the bottom into a collection vessel placed under the conical spout. Finally, the main ice compartment was sheathed in another ice-filled space, which created a jacket of insulation around it, cutting it off from any outside influence once the lid was firmly shut. If any heat was going to melt the ice, it would have to come from inside the device.

The work of Joseph Black foretold that ice absorbed a fixed amount of heat as it melted, so the quantity of meltwater produced was a direct measurement of the heat being released inside. The calorimeter was fitted with an electrical ignition system to spark the contents when everything was in place.

Lavoisier and Laplace performed many experiments with the calorimeter. One of the first involved a guinea pig – something of an exotic creature in those days. Firstly, the pair measured how much carbon dioxide the animal breathed over a fixed period inside a glass bell jar. They then calculated how much charcoal would produce the same amount of gas when burned. To test their prediction, they entombed the guinea pig among the calorimeter's ice for the same fixed time period. Its body heat melted the ice just like any inanimate subject placed inside. Then the charcoal fragment was introduced and set alight. The heat it produced, as represented by meltwater, was a close match to the amount given out by the guinea pig, now fortunately being allowed to warm up after its ordeal in the cold and dark. While it was already known that some of the oxygen we breathe in is replaced by the carbon dioxide we exhale, this ice calorimeter experiment proved beyond doubt that animals (and all life) used a form of combustion to stay alive.

The main job of the calorimeter was to measure the heat released by burning as many simple substances as possible, including hydrogen, carbon (as charcoal) and phosphorus. Lavoisier presented the results in terms of pounds of water melted per pound of substance burned. According to these findings, a pound of burning carbon melts 96 pounds of ice, phosphorus melts 100 pounds, while hydrogen melts nearly 300. His results were somewhat out in many cases, but he had started something, a means of quantifying heat.

The unwieldy units used by Lavoisier and Laplace were soon replaced by the calorie. One calorie was the heat required to raise the temperature of one gram of water by one degree Celsius. France had gone metric by this point. Lavoisier had been a key figure in the metric system's conception but was not around to see it grow up.*

Whatever the units used around the turn of the nineteenth century, the consensus was that they were measuring a physical material, caloric. However, caloric was a special kind of substance, along with light – some also added electricity. Being an element, caloric could not be created or destroyed. There was a fixed amount of it in the Universe and it was perpetually on the move from one area to another. Lavoisier subscribed to the belief that it was made up of indestructible particles, or atoms,

*It should be noted that the Calorie unit used to denote the energy content of food is actually a kilocalorie, in other words a thousand times the size of a calorie with a small c. So if you put an average banana, which contains 100 Calories, in a calorimeter it would produce enough heat to raise 2kg of water by 50°C. And that is just a mere twentieth of what the adult human body needs to run for 24 hours; a full day's food would be an explosive cocktail.

borrowing a theory proposed by Newton that light was made up of the same kind of particle. Newton's great achievement was to show that motion was governed by a set of simple laws that related forces with mass and acceleration. He saw no reason why the corpuscles, or atoms, of light (or heat) did not bounce around according to the same system, like a mind-boggling array of billiard balls.

The atoms of caloric were too small to observe directly, and when working together on a scale large enough to study, they were described as a 'subtle, self-repulsive' fluid. As fluids go, caloric appeared to be a rather gooey one, tending to creep in slow slips and slides, although it could form a mighty blast if conditions allowed. It was believed that caloric attached itself to other particles and only when it was released into a pure state did it create the sensation of heat. The caloric fluid was 'subtle' because it did not appear to have mass, nor did it interact with other materials in the normal way; it only made them hot. It was 'self-repulsive' because it spread out from hot areas, where it was densely packed, to cool areas where it was not. When it was evenly spread it took on a thermal equilibrium. Did that mean the caloric became stationary, or did it keep on flowing in a myriad random swirls that added up to no change over all? Some kind of answer would be provided not by a headstrong – and soon to be headless – scientific aristocrat, but by a school teacher with an interest in the weather.

The English are famous for taking an interest in the weather, but John Dalton took it to a whole other level. At the age of 20 he began recording the weather conditions every day in his diary, taking a note of the temperature and pressure. He carried on the practice for the remaining 57 years of his life, amassing 200,000 observations. Dalton lived in the north-west of England, mostly in Salford near Manchester, where the weather is notoriously changeable, so it would have kept him busy.

As the eighteenth century drew to a close, Dalton wondered what exactly it was he was measuring as air pressure. By now it was common knowledge that air was a combination of gases, at the very least about one-fifth oxygen and four-fifths nitrogen. Water vapour and carbon dioxide were present in small amounts and there was also a mysterious 1 per cent that could not be identified (this was shown to be argon in 1894).

Dalton questioned how a mixture of gases created a single pressure. By 1803, he had developed an answer thanks to the publication of the second and third gas laws the year before by Joseph Louis Gay-Lussac. Both of these relate how temperature, the 'sensible heat' content, of a gas affects its volume and pressure. A gas expands when heated, if allowed. If it is not allowed to expand then its pressure rises. This is all pretty intuitive stuff, but the gas laws also raise questions about how temperature is affected by changes to gas pressure and volume. Does squeezing a gas into a smaller volume make it hotter or only raise its pressure?

Weatherman Dalton came up with his own gas law, the law of partial pressures. Through a series of experiments he was able to show that an atmospheric pressure of 1 was produced by a four-fifths (0.8) partial pressure of nitrogen and a one-fifth partial pressure of oxygen (0.2). Dalton's Law worked in theory for all mixtures of gases – as long as they were not reacting with one another, of course.

That got Dalton thinking about how the gas mixtures were creating the pressure. Gas was seen as an elastic fluid that would fill up any space, and caloric was assumed to be something like it. A gas pressure was the result of the fluid pushing against a solid, immovable surface. In the case of air, therefore, nitrogen was doing four-fifths of the pushing, while oxygen was doing one-fifth. Dalton realised that for this to work, the materials that were nitrogen and oxygen had to be acting entirely independently of each other. The gas inside a jar was spreading out to fill the available space, and the oxygen and nitrogen both spread out to fill that space evenly. There was no portion made up of oxygen and another composed mainly of nitrogen. This was the first empirical evidence that a gas is made up of a collection of individual units, all free to move. In other words it was made of atoms.

Dalton used this discovery to begin to figure out how these gases and other elements combined with each other to make compound substances, carbon dioxide and water chief among them. That in turn offered a way of 'weighing' the atoms of each element comparative to each other. The atomic weights of oxygen and nitrogen were very similar to each other, but hydrogen's was much lower. Balloons filled with lightweight hydrogen had been used as early flying machines since the 1780s. And these hydrogen aviators were following in the wake of balloonists riding beneath balloons filled with just hot air. Did that mean caloric atoms were lighter still?

In 1811, an Italian physicist added a final piece to the gas law puzzle. Amadeo Avogradro reasoned that every gas atom (or molecule, a combination of atoms) contributed the same force to a total gas pressure. This meant that if he filled a jar with pure oxygen so it was at a particular pressure and temperature, the jar would contain a particular number of oxygen atoms (in fact pure oxygen exists as molecules of paired oxygen atoms, but the point remains the same). Doing the same with hydrogen would create a sample of that gas with the same number of particles in it. The weight of the oxygen in our jar was 16 times heavier than the hydrogen and that is because oxygen atoms are 16 times heavier than hydrogen atoms.

It would take another 50 years (Avogrado was rather ignored), but chemists eventually fell upon Avogadro's Law as a means of calculating atomic weights. This led to the periodic table, that mainstay of school-room chemistry. But that is a different story.

The addition of Avogradro's Law created a concept known as an 'ideal gas'. Such a substance follows all the rules set out by the many laws hewn from the work of the pneumatic chemists. Perhaps it's time to take a break and get a cold drink. You see, despite all that heat and light invested in wrangling the gases, the only reason that a refrigerator works is because no gas is ever really ideal.

❄

The search for caloric and the other subtle elements was becoming a wild-goose chase. The first person to break away from the pack was that most unusual of people, an American count. He was born as simply Benjamin Thompson in 1753 in Woburn, Massachusetts, then one of the 13 American colonies under the British crown. When the American Revolution started in the 1770s, Thompson remained a loyalist. He'd married well and had taken over his wife's family properties in Rumford (now Concord, New Hampshire). It became apparent quite quickly that he was on the wrong side, and following a nasty altercation with the locals, Thompson abandoned his wife (and her property) and slid through the line to join the British forces.

His contribution to the war effort was research into gunpowder, which earned him entry to the Royal Society in London. By 1785 he had moved to Bavaria, where he worked as a military adviser and general expert for the local prince. All this work was paid off with Thompson being ennobled as a count of the Holy Roman Empire. He took the title Count Rumford for his lost lands back home.

Towards the end of his Bavarian career, Rumford took an interest in heat. He had long been an opponent of the caloric theory. Instead his work was inspired by the likes of Herman Boerhaave – as we've already seen, this was the same person who got William Cullen into academic research.

Boerhaave had said that heat was the product of the motion of ordinary particles of matter. There was simply no need to involve a mysterious subtle element.

Boerhaave had died long before, in 1738, and in that same year an idea from an entirely different quarter had been proposed that would later come back to haunt caloric. It came from Daniel Bernoulli, a member of an illustrious Huguenot family which had set up home in Basel, Switzerland, and whose combined achievements in mathematics and engineering are too numerous to mention. Bernouilli formulated a mathematical framework that described how gas pressure could be calculated by imagining the gas as a set of particles that moved around the container. The system also took into account the addition of heat, which increased the speed of the hypothetical particles and increased the resulting pressure. Being a mathematician, Bernoulli was not put off by the lack of evidence of such particles. His kinetic system – a system of motion – was a good way of modelling the behaviour of a gas, and that was surely something of use. However, the champions of caloric theory could explain the same phenomena just as well with their elemental fluid, so Bernoulli's work was seen as a feat of mathematics rather than a description of reality.

Winding forwards again to 1790s Bavaria, Rumford had found something that the caloric theory could not explain, based on his experience of working with artillery. When cannons are being bored out, they get very hot. The big link between heat and motion had always been the warming effect of friction. What Rumford wanted to know was this; according to the caloric theory, the heat of a cannon barrel being ground out is a product of the caloric transferring from the boring tool to the gun. So why does the caloric not run out? The heat just keeps on coming.

To investigate, Rumford carried out an experiment at the Munich arsenal in 1798. He placed a cannonball in a barrel of water, and started to drill a hole in it, using a borer made especially blunt to maximise the effects of friction. Within two and a half hours of boring, the water in the barrel was boiling away due to the heat of the iron ball. Nothing spectacular had happened to the iron cannonball. The shavings were just the same stuff as what remained of the ball.

Rumford said that this idea scotched the caloric theory. The whole point of caloric was that it was conserved. It might move around but the total amount always stayed the same. However, the hot cannonball was receiving an inexhaustible supply of heat.

Nevertheless, Rumford knew not from where. Nor could he explain how heat as motion worked in systems where the heat was being conserved – such as in the experiments carried out by Joseph Black.

Count Rumford is now seen as the first person to demonstrate an equivalence between motion and heat, but he failed to undermine the foundations of the caloric theory and soon moved on to other projects. Chief among them was the Royal Institution, which he helped set up in London at the turn of the nineteenth century. This was to be a scientific club to rival the Royal Society but with a focus on boosting public understanding of science. The Royal Institution became famous for its lectures on the latest research fields. Rumford appointed Humphry Davy as the first lecturer. Among many achievements, Davy had found that one of the nitrous airs investigated by Joseph Priestley was a pleasant intoxicant when breathed in. Renamed 'laughing gas', it became the first inhaled anaesthetic and is still used today. Davy's lectures proved to be just as big a hit, making him the world's first science celebrity. Davy's assistant Michael Faraday – the inventor of electrical generators and motors – took over the lectureship duties in later years and established a Christmas performance for children, a grand tradition that persists to this day.

Faraday would later become a pivotal figure in the science of cold.

In 1804, Rumford married Antoine Lavoisier's widow, Marie-Anne Paulze Lavoisier – his first wife, left behind enemy lines in Massachusetts, had since died. Madame Lavoisier had been Antoine's right-hand woman during his rise, contributing much to his theories of chemistry and caloric. She was also there at his fall; her father was executed alongside her first husband. Marie-Anne avoided the guillotine herself and remained a powerful figure in the French scientific establishment, doing much to salvage the Lavoisier brand and its ownership of caloric theory. The Lavoisier–Rumford pairing proved a hot and fiery coupling, but fundamentally a poor match. They separated within a year, disagreeing on many things, no doubt, not just the truth about caloric.

In the end the caloric theory survived Rumford's cannonball attack almost unscathed. The advancement of Rumford's idea had been finally repulsed when he was asked to explain how friction was carrying the heat of the Sun through the nothingness of space. What was rubbing on what out there?

In 1800, the English astronomer William Herschel had found the beginnings of an answer. He split sunlight into its spectrum of colours using a prism. He found that a thermometer registered the most rapid rise in temperature when it was placed right next to, but not in, the red light. Herschel had found what we now call infrared radiation (meaning 'below red'). It turns out that what we feel as heat can be transferred by this invisible form of radiation. In fact there were many more forms of radiation we cannot see to be discovered – all related to light, which is the only visible kind.

Around the same time another Englishman, Thomas Young, had shown definitively that light behaved like a wave not a stream of particles. Newton's laws of motion were no good here, and the subtle fluid of caloric atoms was becoming increasingly beleaguered. It would, however, have one more champion, Sadi Carnot, the equally beleaguered son of a French traitor.

*

Sadi Carnot was the first child of Lazare Carnot, one of Napoleon's lieutenants during the Hundred Days, the emperor's last gasp at wresting control of France before the final defeat at Waterloo in 1815. Lazare was forced into exile after Napoleon's exit, and his son Sadi – full name Nicolas Léonard Sadi Carnot (Sadi was a pet name taken from a Persian poet) – began to find his own military career going nowhere as well.

Carnot teamed up with Nicolas Clément, a chemistry professor in Paris. It was Clément who had come up with the Calorie unit around this time. The pair became interested in the 'motive power of fire', in other words how heat could create motion. All in all, this was akin to what Rumford had been up to 20 years before, but the American's work was all display and no data.

Carnot's main focus was on steam engines, which were proliferating as Europe industrialised. His hope was to make them more efficient. To do that he stripped away any problems caused by friction and other design flaws and created an imaginary, perfect 'heat engine'. A real steam engine is an external combustion engine. This means that heat is provided by a coal fire outside the moving parts. The heat from the furnace boils water into steam and it is this that does the work. According to Carnot and others, the engine created motion by directing a flow of caloric released by the expanding hot steam that pushed against the solid pistons – which then turned the drive wheel or whatever. Carnot imagined that the flow of caloric was something like a waterfall that powered a water-wheel. Just as the scale of a water-wheel's motion is dependent on how much water is falling on it and from how high, Carnot reasoned that the motion of an engine was entirely governed by the quantity of caloric moving through it.

For the caloric to move through the piston, it needed to flow from a hot source to a cooler sink. In the context of a steam engine the source is the boiler and the sink is the condenser, where the steam cooled down and turned back into water. The larger the temperature difference between the source and the sink, the greater the flow of caloric and the greater the motive power of the engine. It did not really matter what you used to carry the caloric – steam, water, air, they all did the same thing. The only really important factor was the temperature difference that kept the caloric on the move.

Carnot's assertion was that the motion of caloric was transferred to the motion of the engine, but the quantity of caloric remained the same. Its heat content was not diminished as it made the engine turn. Instead the engine worked simply because heat was being added to the source by the fuel and it was being lost by the cool sink, dissipating into the surroundings.

Carnot died young at just 36, weakened by 'brain fever' and sent to his grave by cholera. It appears that towards the end of his short life he was beginning to distrust the caloric theory and planned to experiment more with the kinetic theory of heat – the heat created by the motion of particles teaming inside a substance.

Ten years after Carnot's death, it would be shown that the conservation of caloric was not the defining factor of the 'heat engine'. Instead it was the conservation of energy, of which heat was just one form. And this would open the door to the workings of a 'heat pump'. This is a heat engine running in reverse that lies at the heart of the fridge (well at the back) and uses motion to push heat, not heat to push motion.

*

While the epicentre of our story returns to Salford, England, where John Dalton has been tutoring the son of a local brewing magnate, we will take a short trip to the East Indies, on board a Dutch ship heading for Batavia (known today as Jakarta).

The ship's doctor is Julius von Mayer, sometimes known by his middle name, Robert. The year is 1840, and Mayer, a native German, has recently completed training as a medical doctor. There is some suggestion that he has taken work on the Dutch East India Company's ship to escape the impact of a scandalous period during his student years.

In those days, Germany was a loose confederation of states rather than the European behemoth that it was to become. Politics at all scales of society was a delicate business, with frequent flashes of disagreement between competing factions. Conventionality was the wisest course of action, but in 1837 Mayer showed a rebellious side when he attended university balls in 'unseemly clothing.' He was not being indecent, it appears; rather, he had joined a banned fraternity group that turned up at social functions in matching outfits. This kind of hell-raising was enough to get a year-long suspension from his course and a short spell in prison. So when the time came to begin a career in medicine, Mayer opted for an adventure overseas.

The voyage to Java took Mayer and other assorted crewmen from the chill of the North Sea to the roasting humidity of the South Seas. On the way, it is said that Mayer had something of a scientific epiphany, which converted him from physician to physics pioneer, especially the physics of energy. He returned to Europe with the kernel of an idea – all motion, all heat, all chemical activity and indeed all biological activity was a product of energy. This energy could not be created or destroyed, but it was being converted from one form to another.

It is not certain what inspired this realisation. Mayer was undoubtedly aware of earlier work on the subject and it is possible he was already tinkering with the idea as he boarded the ship. Mayer is said to have noted that when a horse-powered stirrer was being used to mix wood pulp, the slurry of fibres warmed up. It is likely he would have seen this as a more workaday version of Rumford's cannonball experiment. Other evidence arrived during the voyage. Firstly, Mayer found that the waters tossed by tropical storms were warmer than those of the previously calm ocean. His reasoning was that the winds that caused the waves were transferring their motion energy to the water, throwing them up into towering, surging peaks, but that motion energy was also adding heat, causing the water to warm up.

That's quite a big leap of imagination, and it may have come on the back of a second, more common account of how he arrived at his final conclusion. As ship's doctor, Mayer was frequently tasked with stitching up bleeding seamen, and he had an inkling that the men's blood had changed colour slightly as the ship moved from the temperate Atlantic to the tropical Indian Ocean. While practising in Batavia itself, his suspicion received further confirmation.

A simple cut releases venous blood, the blood in veins that is travelling back to the heart and lungs. This blood has been used by the body and needs to dump its carbon dioxide and take on more oxygen to further feed the vital fire. Mayer knew this was part of the process, hinted at by many and proven by Lavoisier, that was keeping a body alive.

Blood that lacks oxygen looks darker than arterial blood that is brimming with the stuff. Arteries take blood freshly stocked with oxygen away from the heart to where it is needed. A wound filling with bright red blood is an indication to medics that it is serious and has gone deep enough to cut an artery. Mayer knew this but found he had to adjust the rule of thumb in hotter climates. Even a nicked thumb oozed brighter blood than he would have expected back in Europe. The obvious reason for this was that the venous blood still had plenty of oxygen in it.

Mayer did not have a full understanding of the blood's uptake of oxygen and discharge of carbon dioxide waste as we do now, but he made the assumption that the bright blood of his tropical patients indicated that they were burning less nutrients to maintain life. The prevailing understanding was that the oxygen was used to burn food, in some unknown, controlled way, to release heat. The heat of this oxidation was then driving some kind of life-producing 'engine' in the body.

Mayer described his thinking this way:

> *One great principle of the physiological theory of combustion is, that under all circumstances the same amount of fuel yields, by its perfect combustion, the same amount of heat; that this law holds good even for vital processes; and that hence the living body, notwithstanding all its enigmas and wonders, is incompetent to generate heat out of nothing.*

Mayer did not stop there but combined his theory with his other observations in considering how the living body was giving out heat to its surroundings. In warm places bodies release a little heat to the surroundings, while in colder places they positively pump it out. The evidence from the blood showed Mayer that the body burned more fuel in cold places than in hot ones. That might seem obvious now, but Mayer used it to reveal something about heat and its energy at large.

'We are driven,' he said, 'to the conclusion that it is the total heat generated within and without that is to be regarded as the true calorific effect of the matter oxidised in the body.'

So cold weather, just like hard work and exercise, required the body to oxidise more food. There is a fixed relation between what the body burns and what the body does, including – this was Mayer's masterstroke – transferring its energy to other objects as motion and heat.

Going back to the horse stirring that wood pulp, nosebag attached, its labours are fuelled by the oxidation of the food it is eating. That process makes its body warm, but also makes its body move, and in turn makes the stirrer move. The energy from the food is being converted into motion in the horse, then transmitted by the motion of the stirrer to create motion in the wood pulp. Moreover, as Mayer had noted, the wood pulp is warmed up as it slowly swirls. The motion energy was being converted into heat energy.

Mayer had figured out the first formulation of what is now known as the First Law of Thermodynamics. This states that energy is conserved; it cannot be created or destroyed but is converted into different types. It was becoming clearer that heat, or the lack of it, was due to energy entering and leaving a substance.

Mayer headed home to Germany in 1841, and set about investigating and promoting his theory. However, he was ignored for many years and missed out on his rightful place in history. The unit of energy used by modern physicists today is named the joule, not the mayer. The praise for Mayer's breakthrough was to be heaped upon another researcher, a fact that sent Mayer into suicidal depressions, drinking binges and a spell in a mental health institution. So who was this Joule person? To find out we return to Salford on the outskirts of Manchester, England, where the Joule family ran a brewery business.

*

Just as Julius Mayer had used his medical knowledge as a gateway to the study of thermodynamics, James Prescott Joule used his own particular field of expertise: beer. As Mayer was sailing to the East Indies, Joule was working in the family brewery situated beside the prison in Salford. James Prescott was a third-generation brewer and was born into a family grown wealthy on the beer business. He and his brother were tutored at the family home by the finest minds the north-west had to offer. James Prescott gave up full-time study at 15 and began to help his father, Benjamin, at the brewery. However, he retained a tutor for pleasure and took lessons from none other than John Dalton, the man who had introduced atoms to the modern world.

The Joule brothers were intrigued by electromagnetism, a new field of physics where magnets and electricity had been shown to be inextricably linked phenomena. The boys delighted each other and terrorised the rest of the household by administering electric shocks, using electric generators. A cutting-edge technology, generators had been invented at London's Royal Institution by Michael Faraday the decade before. In 1837, still only 19, James Prescott had begun a serious study of electrical heaters and motors as he sought ways to make the brewery equipment more efficient.

Joule knew that motion was required to generate electricity; a conductor (basically a wire) had to be spun inside a magnetic field for a current to flow. The motion was due to the conductor being given kinetic energy – a fancy way of saying that it was being kept in motion by a force. The faster it was spun, the larger the electrical current produced. The conductor also grew warm, even glowed red hot, as the current flowed through it. Joules made the same connection as Mayer: the kinetic energy of the conductor was being converted into electrical energy in the wire, which in turn was becoming heat, or thermal energy.

The electrical energy is pushing a torrent of electrons, tiny charged particles, through the conductor. This is the 'current'. As the current of electrons travels along, the particles collide with other objects in the material, the atoms. These collisions transfer some of the electrons' energy to the atoms making them wobble around more. This wobble, vibration or other motion at the atomic level is thermal energy. In hot substances the atoms simply wobble further and faster than in colder ones.

Joule knew none of this and was relying on the intuition, the hunch, that since motion caused heat, heat must be some kind of internal motion. This is the kinetic theory of heat that was beginning to overwhelm caloric. Joule grew up in an English industrial heartland where ideas about how to harness heat and motion to do work were being pondered by great engineers, many of them personal friends of his tutor, John Dalton. Dalton himself would have been regaling his pupil with the properties of the atomic world on a regular basis.

In an 1843 paper on the subject Joule set out his thinking:

> *It is pretty generally, I believe, taken for granted that the electric forces which are put into play by the magneto-electrical machine possess, throughout the whole circuit, the same calorific properties as currents arising from other sources. And indeed when we consider heat not as a substance, but as a state of vibration, there appears to be no reason why it should not be induced by an action of a simply mechanical character, such, for instance, as is presented in the revolution of a coil of wire before the poles of a permanent magnet.*

To prove it he built a 'magneto-electrical machine' – we'd call it a generator today – that induced a current in a wire encased in a simple calorimeter. The calorimeter contained a pound of water with a coiled wire inside it. The whole thing was free to spin between the poles of two magnets. As it moved through the magnetic field, a current was induced to flow through the wire, which in turn warmed the water. The spinning motion was supplied by a simple but ingenious pulley system where a pound weight was winched up and then left to drop. As it fell, the weight spun the apparatus. Joule's experiment was to count how often he needed to drop the weight to heat the water in the calorimeter by one degree Fahrenheit.

However, Joule wanted to find a more direct link between motion and heat. He even interrupted his honeymoon in the Alps to take the temperature of the water at the top of a waterfall and then again at the bottom. The theory was that some of the kinetic energy of the plunging water would be converted to heat. He only succeeded in getting very wet. The experiment was a failure – but he still managed to make a success of the honeymoon. He and his wife Amelia were to have two children, but Amelia died along with a third in childbirth in 1854. James never remarried.

The waterfall experiment had been mentioned in passing in Joule's explanation of his most famous experiment. It used a similar pulley system to his first one on electrical heating effects but was altogether more straightforward. Instead of using electricity, Joule heated water by spinning a paddle inside the calorimeter. The friction between the rotating paddle and the water pushed the temperature up ever so gradually. Known as Joule's 'second experiment' (although there were dozens more) this was a less muscular take on Count Rumford's theatrical cannonball set-up.

In 1843, Joule presented his total findings at a British Association for the Advancement of Science meeting in Cork, Ireland. Having set out his various methods he declared: 'After reducing the result to the capacity for heat of a pound of water, it appeared that for each degree of heat evolved by the friction of water, a mechanical power equal to that which can raise a weight of 890 lb to the height of one foot had been expended.'

In other words lifting 890 pounds to your knees uses the same amount of energy as heating a small jug of water by one degree. Whether they were startled by its implications or simply baffled by it, the gathered scientists met Joule's proclamation with silence. His paper went on to use his figures to predict that Niagara Falls warms up by one-fifth of a degree during its long fall. Cue more quizzical looks.

Joule had measured the 'mechanical equivalent of heat'. His figures were a bit out but the paradigm shift that lay behind it was one of the biggest the scientific world would ever see. Over the next decade the substance of caloric gradually faded into oblivion, as the old guard accepted the theory. Heat was coming to be understood as a form of energy. And cold was the result of that energy being in short supply.

*

James Prescott Joule was neither a prime physical specimen nor a confident speaker, and he must have been used to his public lectures being less than successful. In 1847, he gave a talk in Oxford – and was asked to keep it short because few in the audience could master what he was on about. However, this time it was different. The meeting was attended by William Thomson, a fiendish intellect, who at barely 23 was already the professor of natural philosophy at the University of Glasgow, the alma mater of Cullen and Black.

Thomson, later to be ennobled as Lord Kelvin, was one of the few to grasp the implications of Joule's work, and the pair would eventually become research partners. However, even Thomson remained a supporter of the caloric theory for a few years yet, due to a commitment to caloric that arose from his admiration of Carnot's perfect heat engine.

Thomson's interest was pricked by what he thought was a flaw in Carnot's idea – or perhaps a source of eternal power. Along with his older brother James, who was an engineer, Thomson realised that a heat engine powered by ice might be used as a perpetual motion machine.

Admittedly an ice engine would be a slow-moving device, but one capable of immense power. The basic idea was to use the expansion of ice as it froze to push against a piston. The melting ice would then contract, and the piston would fall back. Joseph Black had shown that the temperature of water did not drop as it froze into ice. Therefore, an ice engine could do work and create motion without a change in temperature. That was not possible in a Carnot engine. For the cycle to work, there needed to be a temperature difference making heat flow through the engine. Without a temperature difference the ice engine would work away without needing to add any heat to it – it would be a perpetual motion machine.

It was James Thomson who made the breakthrough that solved this conundrum. For the ice engine to do work, to create an up-and-down motion in the piston, it had to push against something. This was the weight of mechanical parts of the machine, which pushed down on the water or ice, increasing the pressure bearing down on it. James predicted that the increase in pressure would reduce the freezing point of water. So liquid water being squeezed by the engine would have to be much colder to freeze. As it did freeze, it pushed up on the piston. Disengaging the piston reduced the pressure on the ice, and allowed the freezing point to rise to the normal level. Therefore the temperature of the engine would have to rise before the ice melted. As the ice became water again, it contracted and the piston dropped again, raising the pressure and dropping the freezing point. When all that is said and done, such an engine does require a temperature change to work. And thus the Thomsons had solved the enigma of the perpetual ice engine.

Back in late-1840s Scotland, William Thomson had moved on to thinking about how pressure and temperature were linked in gases. He was looking for a universal unit of temperature that was not dependent on devices calibrated to arbitrary points. The gas laws told him, as they had told others before him, that reducing the temperature of gas also reduces its pressure. The reductions are proportional to each other, so when gas pressures are measured in a laboratory, they can be presented as a nice straight line on a graph. Each gas presents its own line as it cools, but Thomson found that when they were extended to where the value for pressure became zero, they all converged on a single point: -273.15°C. This is absolute zero, the temperature at which no more heat can be lost – there is no caloric left in the substance as

Thomson would have understood it then. In 1892 Thomson became the First Baron Kelvin, so the temperature is also described as 0K (note that there no degree symbol used here, and the word for the unit, 'kelvin', takes a small 'k'). The kelvin scale uses the same increments as the Celsius one, so water freezes at around 273K and it boils at 373K.

In Thomson's day investigations into matter at or close to 0K were not possible – there were no refrigeration systems capable of anything like these temperatures. However, the discovery did start a race to achieve this incredible cold, if possible, and in so doing it would provide some brain-boggling insights into the nature of matter itself, such as liquefied gases, magnetic oxygen, liquids that flow uphill and levitating superconductors.

But before then, the world would need a refrigerator, and it was Joule and Thomson working together who found the mechanism that made it possible.

William Thomson and James Prescott Joule joined forces in the mid-1850s. By this time Thomson had got on board with the kinetic theory of energy doggedly expounded by Joule throughout the preceding decade. They made an odd couple. Thomson was a maverick academic with an aptitude for mathematics. Joule was a quiet gentleman scientist with no formal education – he had three degrees by his death but they were all honorary. Joule and Thomson seldom met. Instead they collaborated by letter, with Joule doing the experiments and Thomson doing the heavy lifting when it came to the analysis of results.

The many experiments performed by Joule included an investigation of free expansion. He was not the first to do this but the process is dubbed Joule expansion nevertheless. The purpose of the experiment was to show that gas volume and gas pressure are inversely proportional – when one goes up, the other goes down.

The process involves two containers, identical in every way. They are connected by a tube that can be closed to isolate one container from the other. One container is evacuated so that it holds a vacuum. The other is pumped full of a gas at some known pressure. Then the partition isolating the two containers is removed, and the gas is left to expand and fill both of them. The gas laws suggest that because the volume has doubled, the gas pressure should halve. However, the temperature of the gas, which is really a measure of how fast its particles are all zinging around, should stay the same.

Joule found that this was the case – almost. He noticed a tiny drop in temperature and wanted to find out why this occurred. The mathematics got a little complicated here and Thomson was able to help out. It turns out that Joule expansion is a more or less theoretical process. No gas actually obeys the gas laws completely. The gas laws relate to an ideal gas and most real gases come so close to this ideal state in normal conditions that the tiny inconsistencies do not really matter.

Nevertheless the inconsistencies are there. Joule and Thomson found a way to capitalise on them in order to create large drops in temperature with little more than a nozzle and a pump. The effect they discovered is aptly known as Joule-Thomson expansion – physicists are nothing if not consistent when it comes to naming things. The set-up is similar to the one used in Joule expansion, only the connection between the two containers is a much tighter nozzle and the gas is not just left to expand, but is forced from one container into the other.

The result is that the gas expands extremely rapidly to fill the void and as it does so its temperature plunges. The drop in temperature means that the gas particles have slowed down, and the reason for that slowdown is reality getting in the way of the ideal.

The gas laws do not take into account forces acting between the gas particles. They are supposed to just bounce around, colliding off each other and the surroundings.

However, there are very small but not inconsequential forces that are constantly pulling gas particles together. So when the gas is forced at high pressure into an empty space it has to work hard to fill the space – the work is done by each particle breaking free of the forces that pull them together. That all takes a bit of energy, so the particles slow down and the gas gets cold.

Joule–Thomson expansion, discovered in 1852, is the effect behind our refrigerators. Science and engineering had found a way to create cold. But few people cared. While researchers were busy decoding the secrets of energy, businessmen were busy getting rich by transporting millions of tonnes of icy slices from the lands of winter to the permanent summer of the tropics. Frozen water had become a global commodity.

SEVEN

An Ice King – Or Two

The sweltering inhabitants of Charleston and New Orleans, of Madras and Bombay and Calcutta, drink at my well … The pure Walden water is mingled with the sacred water of the Ganges.

Henry David Thoreau, Walden Pond, 1847

Behind the magic and machinery of the long quest to plumb the depths of cold and harness its nemesis heat, natural ice was there all along. In 1660, Charles II of England – the grandson of James I, England's first air-conditioned monarch – had a new pleasure garden built in central London. The result was a band of royal soil that connected the two already existing Royal Parks – Hyde Park and St James's Park. Today, Charles's garden is known as Green Park, an epicentre of sandwiches and sunbathing on the edge of the city's West End, but Charles's original vision was altogether more lavish. It was a place where he could entertain dignitaries, foreign and domestic, in the style he had come to admire while in European exile during the civil wars and Oliver Cromwell's Commonwealth that followed. To that end, he claimed for the Crown the meadowland north of Westminster, previously used largely as a duelling ground and for the odd rebellion, and had it walled off and stocked with deer. Charles then added an ornamental canal and a hunting lodge – plus an icehouse, 'as the mode is in some parts in France and Italy and other hot Countrys, for to Coole wines and other drinks for the sumer season'.

This royal icehouse, completed in October 1660, was the first of its kind in Britain. In the end the site had six icehouses, all built in Sandpit Field just beyond the eastern wall of the park (named Upper St James's Park at the time).

Sandpit Field is now covered by the far end of St James's Place, and any remains of the icehouses are under some of London's most well-appointed properties, where one must assume that the refrigeration available is still very much state of the art. To echo the Mari Tablet where our story began, Charles's icehouse 'which never before had any king built' caused a sensation among English society.

As we have seen, Robert Boyle was enthralled enough with the fad for cold in the 1660s to start a new science. Edmund Waller, a politician well versed in sycophancy, just wrote a poem:

On St. James's Park as lately improved by his Majesty

Yonder the harvest of cold months laid up,
Gives a fresh coolnesse to the Royal Cup;
There Ice like Christal, firm and never lost,
Tempers hot July with Decembers frost,
Winters dark prison; whence he cannot flie,
Though the warm Spring, his enemy grows nigh.
Strange! that extreames should thus preserve the snow,
High on the Alpes, and in deep Caves below.

But the Mari Tablet was written 1,900 years before Charles was restored to the throne, and he and his icehouse were somewhat late to the party, no matter how cool the royal cup.

Ice, that commodity of the ancient world, that marvellous plaything of the Renaissance, had been harvested and stowed away in more venerable icehouses for centuries. Come the seventeenth century, the yakhchals were still churning out fresh ice in the deserts of Persia. The sharbats of Constantinople's merchants were cooled with ice chips brought down from Mount Uludag. This peak, located across the Sea of Marmara, is so lofty that it was long regarded as a local Mount Olympus, a seat of the gods, by the ancient Hellenistic Mysian culture that once ruled the area. To the west, Sicilians enjoyed sorbets made from the snows of fiery Mount Etna, some of which made it all the way to Rome. In southern Spain hard-packed winter snow from the Sierra Nevada was stored in large ice pits up on the mountainsides. In summer, blocks of snow were moved in nightly mule trains along extensive trading routes established around Granada by the Moors, and all thanks to their importation of the Middle Eastern icehouse technology.

Now it was the English who were importing the same technology, thanks in the main to aristocrats returning from the exile imposed on them by the revolutionary turmoil that rocked Britain and Ireland in the middle of the seventeenth century. As order – and the monarchy – were restored, the gentry attended to their country estates and it became de rigeur to include an icehouse somewhere near the mansion.

In keeping with the wisdom of the Persians, icehouses were invariably sunk deep into the ground, where it was assumed that it was colder than on the surface. Above ground the ice store was sometimes topped with an ornate folly in the fashion of the day, or simply a thatched pyramid – some were even built inside the towers of a castle. More frequently they were visible only as an earth-worked mound with a door leading to the chamber beneath.

It is said that as a child, the author J. R. R. Tolkien played in Moseley Park in suburban Birmingham. The park had once been part of the estate of a great country house that had since passed into public hands (donated by the Cadbury family, famed English chocolatiers). Tolkien would have found the sunken doorway to the estate's icehouse (since restored) hidden abandoned among the park's trees. It is often wondered whether this was the inspiration for the domed hobbit houses of the Shire in the writer's fantasy novels, *The Hobbit* and *The Lord of the Rings*.

The Moseley icehouse is typical in its design, following the construction rules as set out by Philip Miller in 1768; an icehouse should be built in well-drained land; waterlogged soil will melt the stored ice. The best sites are out in the open, where sun and wind are able to dry the soil; locations secluded by woodland are invariably too damp. Where no proper drainage is available, the icehouse must be built above ground, but buried in soil or shielded in some other way nevertheless. Inside, the chamber should be round and built to last. Miller suggests that ice may be required to be in the chamber for three years as not every British winter brings a deep enough frost to restock the store.

In an attempt to seal off the ice from the outside world, the chamber was approached through a tunnel, sometimes making twists and turns to block any blasts of warm air from outside when the outer door was opened. In most cases there was at least one internal door, which would have been edged in leather to create a tight seal when pulled shut.

Having negotiated the tunnels, steps and doors, the iceman would have reached the storage chamber itself. The ceiling above was a brick dome, and the ice was packed into a chamber under the floor and accessible through a hatch. Earlier designs incorporated chambers that tapered in at the bottom and were even full cones. The idea here was that water trickled down to the bottom and out through a drain where it could not 'rot' the solid ice. The ice was loaded onto a grill or timber base raised above the deep floor to keep it out of the pooling meltwater.

Chambers evolved into bell or bulb-shaped constructions, with a double-skin wall, where an air-filled void between the inner and outer brickwork provided extra insulation. The most advanced designs collected the meltwater in a sump, which, if kept clean enough, was pumped out as a source of chilled water.

Building an icehouse required skilled bricklayers, often imported from the Netherlands. Within a few years, the English nobility had imported a whole new royal family from the Netherlands, in a plan hatched in one of those newly built icehouses. The coup d' etat plotters were unhappy with James II, Charles's heir, and looked for a safe, secluded place to make their cold calculations. The treacherous gang chose to meet within the thick walls of an icehouse at the Charborough Park estate in Dorset, England, in the summer of 1686. Safe from the king's spies, they drew up a plan to approach the Dutch nobleman, William of Orange, and offer him the throne. William's small invasion force landed unhindered two years later. The coup was a success, and became known as 'The Glorious Revolution.'

In common with King Charles's Green Park, the grand estates of England were generally many miles from the nearest mountain – and mostly they were small with a scant supply of snow when compared to the monoliths of the Continent. Instead, the English took another leaf from the Persian iceman's handbook (with the help of Dutch knowhow on controlling water). Icehouses came with an associated shallow freezing pool. We may recognise some of these pools today as ornamental ponds, but at the time, if the winter was cold enough, they would be in the service of the icehouse and cleared of their ice once they froze over. Boating and fishing lakes built by damming a small river are another feature of the English country estate. Some of them retain sunken walls, which would have been used as dams to seal off and becalm a section of the lake when freezing conditions prevailed.

Not every winter resulted in an ice harvest, and ice was valuable stuff. The grandest of icehouses came equipped with a little room for the icemen, who watched over the store and dug out the next batch at the whim of their masters. Hampton Court Palace, originally built in 1514 by Cardinal Wolsey on the Thames south-west of London, has an especially grand icehouse. In the 1690s, the new king, William III, rebooted grand plans to convert the palace into London's equivalent of Versailles, albeit in the English style (or was it Dutch?). Anyway, it was only ever really half finished, but impressive nevertheless.

Hampton Court's ice house had initially been a snow well, a brick-lined pit covered with a thatched roof. In the 1690s it was converted into a full-blooded icehouse, eventually enclosed in a 12-sided building covering the original snow well. The well was strengthened with iron bands on the orders of James Frontin, Yeoman of His Majesty's Ice Wells, and enclosed by an inner dome. In the space between the dome and outer wall is a small room fitted out with a fireplace. This was no doubt much welcomed by the icemen on duty inside, but could not have done the ice stashed further in much good.

Such a misplaced feature illustrates how the process of freezing and melting was still a mystery, not least to the ice yeoman Frontin, and that icehouse construction was as heavily influenced by tradition as the need for efficiency. There are many records of failed icehouse projects, despite the builders following all the rules.

Nevertheless, by the turn of nineteenth century, the same kind of hermetic, subterranean icehouses were being built by a new gentry across the Atlantic in the United States. Even George Washington wrote about his stores of ice in 1785. One Boston Brahmin family, the Tudors, had an icehouse at their country estate near what is now Saugus, Massachusetts, brimming with ice collected in the sharp New England winter. The Tudors took pleasure in ice just as King Charles II had done. However, Frederic Tudor, the middle son of the family, would be instrumental in making ice a household necessity not a luxury and transporting it to markets across the world. Today, it is Tudor who is remembered as the true 'Ice King of the World'.

❄

Ice in the north-eastern US states is seldom in short supply due to it being on the opposite side of the Atlantic weather system that keeps Western Europe's winter relatively mild.

By contrast winters along the Eastern Seaboard readily freeze lakes and rivers as far south as Virginia.

Frederic Tudor was not the first Yankee businessman who saw profit in nature's unused, undervalued bounty. In 1802, the Maryland farmer Thomas Moore invented an icebox, which he used to transport his butter to market. Moore's cold, hard butter fetched a better price than the sloppy, rancid products of his competitors. However, Moore saw a better return in the icebox market than the butter one. The problem was that there wasn't really an icebox market, not yet at any rate.

In 1803 Moore received a patent for his device, which he named the Refrigeratory (this was a term borrowed from the cooling coil used in stills to condense liquor). The patent was signed by none other than President Thomas Jefferson, who was so impressed with the icebox that he bought one himself. The design of the Refrigeratory was not that impressive. It was an oval tub made from cedar wood, fitted with a rectangular tin box. Crushed ice and snow was packed into the spaces between the two so as to keep whatever was in the box cool. To stop the ice from melting, the entire chest was wrapped in a cloth lined with rabbit fur.

Moore certainly saw the great potential of his invention, setting this out in an essay on its many applications published in Baltimore shortly after the patent award:

> *The following are some of the useful purposes to which the machine may be applied … Every housekeeper may have one in his cellar, in which, by the daily use of a few pounds of ice, fresh provisions may be preserved, butter hardened, milk, or any other liquid preserved at any desired temperature; small handsome ones may be constructed for table use, in which liquids, or any kind of provisions may be rendered agreeable, as far as it is possible for cooling to have that effect.*
>
> *Butchers, or dealers in fresh provisions may in one of these machines, preserve their unsold meat without salting, with as much certainty as in cold weather; and I have no doubt, but by the use of*

> *them, fresh fish may be brought from any part of the Chesapeake Bay, in the hottest weather and delivered at Baltimore market in as good condition as in the winter season.*

Moore's Refrigeratory – soon rebranded as the less cumbersome Refrigerator – could be made in numbers to any specification. There was certainly no shortage of ice, cedar wood and rabbit fur. However, Moore found a shortage of customers (President Jefferson notwithstanding).

Moore soon turned his attentions to other projects. His brush with the chief of state helped him win major government contracts, and he worked on the roads, waterways and bridges needed by the young nation. His Refrigerator business never got off the ground. The post-revolutionary American family had no need for it.

In 1805, just as Moore went cold on his icebox idea, the 23-year-old Frederic Tudor was in the early stages of his own ice-based business. Moore had shown his cooling idea worked, but nobody wanted to buy it. Tudor's plan was to sell cold to people who did want to buy it, but nobody was prepared to believe his idea could work. It would take him 25 years, a few spells in prison, unimaginable debts, a dose of yellow fever and a mental breakdown, but in the end he proved those early doubters wrong.

The family icehouse at the Tudor's Rockwood estate was a luxury. Frederic and his brothers and sisters had cool drinks all summer and ate home-made ice creams. Their idyllic childhood was supported by their father William, a Harvard-trained lawyer and judge, and a family fortune amassed by grandfather John Tudor, a baker and businessman, who'd arrived in Boston from England as a six-year-old with his penniless mother.

However, wealth could not fend off tragedy, and in the winter of 1801, the Tudor's second son, John Henry, became an invalid, probably suffering from a protracted case of tuberculosis that had spread to the bones. It was decided that the 19-year-old would take a trip to a warmer climate, and would be accompanied by his younger brother Frederic. Frederic, who was only 17 but had already dropped out of school and had been working as an apprentice to a Boston merchant for nearly four years, was thrilled at the opportunity to see more of the world.

The pair set sail for Cuba, hoping to dabble in the sugar and coffee trade with the $1,000 their father had given them for the trip. Neither boy enjoyed the voyage south, complaining of seasickness and sunburn. Arriving in Havana in spring, they perked up and enjoyed themselves for a while, touring the island, which was still a Spanish colony in those days, all the while making ill-considered business deals that invariably ended in a loss. However, when the febrile heat of the Caribbean summer set in, bringing with it the threat of yellow fever, the brothers could not stand it any longer. They took passage to Charleston in the Carolinas, as all the while John Henry's condition grew worse. The climate of Charleston was just as intolerable for the New England boys, so they pressed north, finally settling on a spa in Pennsylvania, where John Henry could recuperate. But to no avail. The boy was worsening and was taken to Philadelphia by his mother, who travelled south with her eldest son William. A few months after Frederic and William had got back home to Rockwood, the news arrived that John Henry, too ill to travel further, had died in Philadelphia.

Frederic mourned the loss deeply, and the experience of his first voyage to the torrid South proved formative. Sweltering in the Havana heat, he must have thought of the cooling ice he enjoyed back at home and of the relief it would have given his dying brother.

After two more years working for a merchant friend of his father, he was ready to start a business of his own. His first venture was a family affair, investing with his father, known by his sons as the Judge, and brother William in property – seen as a sure thing in a city growing as fast as Boston was.

By 1805, Frederic had hit upon his life's work – it too would be a family business, going all in with William. Frederic began a diary on the first day of August. The cover is inscribed 'Ice House Diary' and the first entry reads:

> *William and myself have this day determined to get together what property we have and embark in the undertaking of carrying ice to the West Indies the ensuing winter.*

The plan was simple enough – overly so, it would turn out. Point one: approach the colonial authorities of the Caribbean to secure monopolies for the supply of New England's ice to their respective islands. Point two: cut ice from the pond at Rockwood in winter and transport it to Boston harbour. Point three: transport the ice south on a Bostonian brig (details to be confirmed). Point four: sell the ice and become fabulously wealthy. Point five: repeat the following year.

The overarching logic was perfect. Tudor would take something that was free and plentiful in one place and sell it where it had some value. Over the next 30 years, he would discover time and time again that his ice-trading scheme was less than perfect in practice.

At an unknown date, years after his first foray as an ice trader, Tudor emblazoned the cover of the Ice House Diary with a nineteenth-century self-help mantra: 'He who gives back at the first repulse and without striking the second blow despairs of success, has never been, is not, and never will be a hero in war, love, or business.'

It would seem that things had not been going well.

However, Tudor was seldom short of optimism and self-belief and these were in full evidence when he launched his ice business. While family friends politely declined to invest, the hearty and heartless merchants of Boston at large thought the idea hilarious and ridiculous in equal measure. The Tudor brothers pressed ahead. William and a cousin called James set sail for Martinique, their mission being to butter up the island's governor and get that all-important monopoly. Although they laughed at him now, Tudor felt sure the bosses of the Bostonian trading fleet would soon try to get in on the act once they saw the easy profits.

Frederic stayed at home to organise the harvest of ice and its haulage south the following spring. William and James, meanwhile, were having mixed fortunes. The passage south was hampered by bad weather, pirates and a skirmish between unidentified European navies. Upon arrival in St Pierre, Martinique's charming, but hot, capital, the young men began to move in the right circles to get access to the Prefect, the French governor. That took more than a month, and their request for a monopoly was met with two preconditions: firstly, the Prefect wanted to know how the ice would be sold, and secondly he wanted $400 up front.

The first condition was tricky enough. It appears that the Tudor boys had not planned much beyond getting the ice to the island. The second one was also a problem. They did not have that kind of money to spare, so they simply bribed the Prefect with a couple of gold coins. The monopoly was theirs.

Meanwhile back in Boston, Frederic had problems of his own. He could not find a captain willing to lease his ship for the first ice transport. A sail-powered cargo ship had to be loaded with care to balance the vessel, and Boston skippers were used to carrying a range of cargoes. Filling the hold with ice would be folly, they argued. The ice would melt, making the hold awash and damaging the other cargo. And as the water leaked out of the hull, the ship would become lighter and sit dangerously high in the water.

Frederic threw money at the problem. He bought his own ship, the *Favorite* – and with it went any hope of a profit from the first run to Martinique.

On the island, James had been struck down with yellow fever, a common disease in those parts back then. Leaving James to battle the disease, William toured neighbouring islands on the pretext of evaluating them as future markets for Tudor ice. His island hopping had little impact, and he made no preparations for the arrival of Frederic and his ice at St Pierre. To be fair he had no way of knowing if his brother was even coming.

But coming he was. Loading of the *Favorite* had finished by mid-February. Her hold had been lined with boards to offer a layer of insulation. Rough-cut hunks of ice, smashed out from the pond at Rockwood, were hauled in wagons the nine miles to the city docks, and nestled into the ship before being sealed with a thick layer of hay. The *Favorite*'s illustrious departure was recorded in the *Boston Gazette*, although in less than illustrious terms:

> *No joke. A vessel with a cargo of 80 tons of Ice has cleared out from this port for Martinique. We hope this will not prove to be a slippery speculation.*

We could imagine Frederic's contented smile as his ship headed for the open sea – his hold held 130 tons of ice, not 80. Pure, frozen profit.

His smile would have begun to waver as he made landfall in Martinique after a fast voyage free of incident. James had left to join William's Caribbean jaunt, and in their place stood an island agent holding the monopoly documentation and a note from Frederic's brother. The note's message was blunt: there was no market for ice in Martinique, and Frederic shouldn't bother staying long. William had left no hint about where he was, or where else Frederic should try his luck.

William had made no attempt to prepare a place to store the ice. Frederic appears to have been under the impression that this was one of the main reasons for travelling to the island in the first place. At the time it was still assumed that an effective icehouse had to be sunk into the ground, but Frederic would later pioneer timber-built icehouses that worked just as well as the elaborate snow wells and ice pits of the Old World (and were a lot cheaper to erect). However, in March 1806 he had a ship full of ice (a large portion appears to have remained frozen on the three-week voyage), with nowhere to store it. And, as would become clear soon enough, he did not really have anyone to sell it to either.

Frederic resolved to sell the ice on the dockside. He is reported to have been offered $4,000 for the whole lot, but that would not even cover the cost of the ship, so he turned it down. At 16¢ a pound, the ice trade was not brisk. To the island-born locals it was a novelty, but few had any idea what to do with it. Frederic reported that customers were baffled when their purchase turned to water when left in the sun or placed in the bath to keep cool. Frederic handed out instructions with each purchase, advising customers to wrap the ice in blankets, but he became alarmed by the lack of interest in his product, rapidly melting on board the *Favorite*.

A glimmer of hope came in the form of ice cream. St Pierre had a pleasure garden, and its owner welcomed the idea of selling his patrons a cooling treat. But he doubted that it was possible. Ice cream still retained its mystique in this part of the world, but Frederic knew better.

Ice cream is not made by adding ice to cream. Instead Frederic mixed his ice with salt water. This was the frigorific mixture of Renaissance wizards, this time deployed on a volcanic island just north of the Equator. Making ice cream in those days was a laborious business, and Frederic had his work cut out – he is reported to have used 60 pounds of ice. The cream and sugar were placed in a metal vessel that was then plunged into the frigorific mixture. As it was cooled by the icy liquid, the cream had to be stirred constantly. Frederic spent the best part of the day preparing the ice creams, which fetched $300 in St Pierre's park the following evening.

The remarkable appearance of ice cream even made the island's newspaper.

Although his ice was melting faster than he could sell it, Frederic managed to rack up a turnover of $2,000, before heading back to Boston in March with a cargo of sugar to boost the coffers. The return voyage was a further blow to Frederic's fortunes. The *Favorite*'s masts were lost in the storm, requiring costly repairs back at St Pierre. The ship finally slid into Boston Harbor at the end of April.

William and James were nowhere to be seen, leaving Frederic to count his loses and keep his fury on a simmer. The hapless pair finally walked ashore in June, having been robbed by Spanish pirates, got lost at sea and finally quarantined for yellow fever. Their fact-finding mission had revealed only one thing. It was possible to win monopolies for ice trading in the British West Indian colonies. But there was a problem. They would have to go to London and ask permission from the Privy Council, an inner court of esteemed advisors, who answered only to the king.

All in all, the first season of the ice business had not been a success. Frederic had used up all his capital and had a $3,000 loss to show for it. Frederic's fury gave way to self-pity as he bemoaned his error in allowing William and James to be involved in such an important part of the venture. Nevertheless, he was going to try again, only this time he would not be relying on anyone else to ensure success.

❁

In 1806, William was packed off to England to attempt to negotiate a trade deal for the West Indian colonies. He was accompanied by his little sister Delia and his parents. They were using the trip as an opportunity to mix with the high society of London – and later Paris – in an attempt to marry off Delia to a wealthy suitor.

It was becoming clear that business acumen was not a family trait. The land deal that Frederic and his father had entered into a few years before was not proving to be quite as sure a thing as they had hoped. Their plots south of Boston were not making anything. It would seem they had bought the only land that no one wanted to build on as the city expanded in all other directions.

Reckless optimism did run in the family, however, with Judge Tudor borrowing heavily to pay for the European jolly. The family fortune was running out, and only Frederic had a plan, thoroughly reckless and optimistic, to make another. But having taken full control of the business, Frederic would find that he was still at the mercy of other forces – this time it was the justice system and international politics.

Learning a lesson from the debacle in 1805 in Martinique, Frederic arranged for an ice house to be built in St Pierre. This would prevent his stock from disappearing faster than he could sell it. The other thing 1805 taught him was that there were not that many customers for ice on tropical islands. His response to that problem was to expand. He built another ice house in Havana in preparation for a supply run in the spring. He felt sure that the demand for ice cream and cool drinks would be higher in the Caribbean's largest and most modern city.

In 1807, the second shipment of New England ice was sent to the Caribbean, this time as stock for the Havana ice house. Frederic did not go with it and instead entrusted its sale to another cousin, William Savage, brother of James. The Cuban ice house seems to have been barely fit for purpose, but William sold $6,000 of ice in two weeks. The ship that had carried the ice down – the *Trident*, the *Favorite* having been long since disposed of – was signed up to bring a cargo of molasses back to Boston. This would have boosted the takings further and put Frederic into profit. However, the customer for the molasses went bust, and Frederic was back in the red. There was no money left to supply Martinique, and the brand new Tudor ice house there was left empty.

With impeccable timing, William Tudor wrote to Frederic with news, both good and bad. The British had granted the Tudors exclusive rights to sell ice in their West Indian colonies. It appears that the authorities had been suspicious that the venture was cover for smugglers, and were minded to turn down the request. However, William managed to convince them by having a doctor testify to the medicinal qualities of ice, which would be valuable, even life-saving, in Jamaica and Barbados. William had surpassed himself, but Frederic had no money (only debts) and so could not open these new markets.

In the winter of 1807, Frederic was preparing to sail to Havana to rebuild the icehouse. Sales had been good, but too much of the ice was melting once it was unloaded. With his ice store half finished, Frederic learned that yet another obstacle had risen between him and a viable ice business. President Jefferson, ironically one of the few who saw a future in ice technology, had put a stop to all shipping to foreign ports, to try to assert the sovereignty of the United States at a time when Britain and France were frequently at war. The idea was that a ban on American exports would be the best way of showing that the United States was a neutral in the conflict, which often involved naval skirmishes close to American waters. The reality of the ban was to make many a Boston merchant go out of business, but not Frederic Tudor – he'd more or less put himself out of business many times over. Nevertheless, there would be no ice runs to Cuba while the embargo was in place.

Frederic returned to New England and nursed his ambitions while avoiding his increasingly belligerent creditors. As the Tudor bad luck would have it, William announced that the family now had exclusive rights to French territories as well as British. They had sewn up the market but were banned from exploiting it.

The trade embargo stayed in place until 1809, lifted after it was possible to harvest ice for a run south that year. Frederic had to wait for the next winter, but as it approached he was arrested for non-payment of debts. He was served papers on State Street, in the heart of Boston's commercial district and in full view of his peers. The family scraped together enough to spare him jail, but the humiliation hit Frederic hard. When the time came for the next shipment of ice to Havana in early 1810, Frederic went with it to escape his tarnished reputation.

The year 1810 was an unusual one for Frederic. He made a profit, although barely enough to dent his debts (however, most of that year's success had been managed by a cousin, Arthur Savage, because Frederic had been struck down with yellow fever). 1811 was a return to the norm. Frederic planned to open a business in Jamaica, and sent his younger brother Harry to be his agent. A schooner, the *Active*, was dispatched to Kingston with the first supply of ice. It sank.

In March 1812, the inevitable happened: Frederic went to debtor's prison. However, after juggling his assets to pay his way out he emerged even more sure of himself. Frederic set up an ice store in Boston to hold stocks that could be sent south throughout the summer. His confidence came from the Havana operation, which was a going concern. Testament to that fact was that other ice traders were attempting to enter the market. Frederic enjoyed watching them endure the same difficulties with handling ice in the tropics that he had suffered – but overcome. The sinking of the *Active*, the trade embargo, the debts, the prison stretch – these were all just bumps in the road. What else could go wrong?

In June 1812, the United States went to war with the British Empire. Any American ship on the high seas was a target for the Royal Navy. Frederic and his melting store of ice were stuck in port yet again.

The war lasted for almost three years. In that time Frederic was jailed twice for his debts and considered selling Rockwood, the treasured family farm – and the source of his ice. However, the proceeds from that and all of the other family property would leave a shortfall of $10,000.

Frederic was at risk of arrest wherever he went. So with the seaways open once more, he slid away to Havana.

❁

Frederic had not given up on ice, and he'd arranged for a shipment to arrive in Cuba in the following spring of 1816. However, on arrival he found he had competition. A Spaniard by the name of Carlo Goberto de Ceta had convinced the authorities in Havana that he had the exclusive rights of ice sales in Cuba (this was easy enough to do because Frederic had won his monopoly through bribes in the first place). De Ceta was proposing to install an ice factory using artificial refrigerant imported from Europe. In an attempt to halt the import of ice, de Ceta even travelled to Boston to besmirch the Tudor name (which was not hard).

William Tudor, now a man of letters and retired from the ice business, wrote to Frederic to update him of the situation. He reassured him that the ice was being loaded without hindrance, and de Ceta would not be a problem. He then added that Harry Tudor, who was now running the Boston end of the business, had mortgaged Rockwood to pay for the shipment. If Frederic blew it this time, the family would lose everything.

Back in Havana, Frederic had won a reprieve that allowed him to trade alongside de Ceta for two years (de Ceta would prove to be nothing but a chancer, even more so than Tudor. He never sold any ice; the best his proposed system could have done would be to chill drinks, and one assumes he had hoped to simply block the sale of the superior Yankee product). Nevertheless, Frederic already had another problem. The governor objected to the large Tudor icehouse being built so close to the harbourside. By the time a site had been agreed, there was just two weeks to get the wooden building up before the ice arrived from Boston. It was still being finished as a slave gang loaded it with ice hauled up from the quay.

The 1816 Havana icehouse had a prefabricated cedar frame, which Frederic brought with him from Boston. It was more or less square, and about eight metres high. The ice was stored inside an inner chamber, with about a metre of space between it and the outer wall. The ice was retrieved from a trapdoor in the roof of the inner structure. This roof formed an upper floor that was accessed by stairs on the outside of the building.

The external roof was not fitted for several days after the ice had been stored. The roof slowed the melt rate, and covering the top of the inner store with sawdust, covered with blankets, reduced it to acceptable levels.

Perhaps by chance rather than good design, the Tudor icehouses were effective despite seeming to break all the traditional rules. Melting was inevitable, but Frederic had a good drainage system – the entire floor was sloped to a central channel at the centre, like an upside-down roof, which removed the water that would otherwise accelerate the melting. Also underground ice stores were easier to insulate from warm air outside, but a sealed chamber was unable to shed the latent heat released as the ice transformed to water. This had the uncanny effect of heating the inside of the ice store. Frederic's quick and cheap structure was better ventilated, and as a compromise between all the competing factors it was a more than adequate solution – one that would be copied (and improved upon) across the world over the ensuing decades.

With the ice safely stored, Frederic just had to find customers. He was firmly of the opinion that once someone had experienced the pleasures of ice, they could not countenance life without it. He later summed it up thus: 'A man who has drank his drinks cold at the same expense for one week can never be persuaded to drink them warm again.'

So Frederic became an ice pusher, offering Havana cafes free samples, and outlandish coolers made from large jars insulated with sawdust and moss. These contained 14 gallons of ice (that weighs 60kg, or 130lb) and had to be hoisted with cranes. Frederic marketed the ice coolers with typical self-centredness: 'Drink Spaniards and be cool that I, who have suffered so much in the cause, may be able to go home and keep myself warm.'

Frederic hated life in Havana and yearned to return to New England. The marketing push had limited success, and most of the sales in Havana were for making ice cream. Nonetheless, trade was brisk and finally, after ten years of turmoil, the Tudor ice operation was running at full swing. Ice harvested in winter was being supplied to Havana from the Boston storehouses throughout the summer.

While Tudor ice was used to cool drinks, Frederic began to experiment with its ability to preserve foods. After some tinkering, he found that he could keep Cuban oranges fresh for a month by packing them in ice. This opened up the prospect of reloading the ships that carried ice south with chilled tropical fruits to sell on in Boston. The first and only attempt was yet another costly disaster. The fruit began to ferment and rot under the heavy insulation, producing a disgusting stench. A few of the fruits were good enough to sell, but Tudor's supply of fruit was not really meeting a demand anyway. There was plenty of fruit available all summer, and in winter when preserved fruit would have found a market, the Tudor's ice was in a New England lake not in Havana. And again, Frederic had lost all of the profits he had accrued from selling ice.

❄

As the 1816 summer drew to a close, Frederic got a tip from his brother William about a possible business partner in

Charleston, South Carolina. Havana had proven that the ice trade worked. The total ice trade had increased almost tenfold from the 130 tons carried in 1806 to 1,200 tons that year. Perhaps it was time to bring ice to the cities of the South. The supply runs would be shorter yet the benefits of ice would be appreciated just as much in the sweltering summers.

Frederic would have been nervous on his return to Boston in October of that year. He was still a man who owed a great deal of money. He wasted no time in making for South Carolina, where he was to meet the putative investor General Thomas Pinckney, an old friend of his father. The problem was that the Tudors were all broke, and had no capital to invest in setting up a Charleston ice house.

He managed to persuade some backers that his product was not an exotic luxury for the wealthy but a necessary and affordable commodity for the masses. In the summer of 1817, an advert was taken out in a Charleston newspaper explaining as much, and announcing the opening of an ice house at Fitzsimon's Wharf.

Families were able to buy monthly supplies at a price 'as low as it was in the northern cities'. The Tudor company's large ice house also sold 'Little Ice Houses'. These were cold-storage containers – and were later to be sold by others under the charming name of cellarette. Functionally, if not by exact design, they were a copy of Thomas Moore's Refrigerators. As well as helping to keep the ice supply frozen for longer, the hope was that the iceboxes would be used for the cold storage of foods – and become an indispensable piece of household equipment.

The arrival of ice in the South did not change American society overnight. The purchase of ice and its transportation home was the responsibility of black slaves (Charleston was still very much the slave capital of America at the time, with more African slaves in the town than whites). Nevertheless the first summer of ice sales was a success. The Havana business, on the other hand, was in disarray, with the icehouse forcibly relocated by the authorities. Back in Boston, Frederic's brother Harry had joined him as a *persona non grata* in the city, pursued by creditors wherever he went, and placing further pressure on the family finances. As was so often the case, the Tudors had cleared the path for a viable ice industry but were in danger of going under just as other players joined the trade.

The winter of 1817 did not help. Freezing conditions often come suddenly in the New England winter, with ponds and rivers freezing overnight. However, in 1817 the cold weather never really came and that meant ice was in short supply. There was just enough for two cargoes to Havana but neither ship made it to its intended destination.* The 1818 profits from Havana would be zero.

Frederic knew that the future lay in supplying ice to Charleston and the neighbouring cities, such as Savannah, Georgia, with the biggest prize being New Orleans at the head of the Mississippi Delta. But just as success beckoned, Frederic had no money for the relaunch and he could not go home for fear of arrest. Failure was beckoning as well.

The irrepressible Frederic was on the verge of quitting, but with a trademark twisting of logic he declared that 'My reputation is now so far pledged that I must advance.' It was ice or nothing. He had made a small profit in Charleston despite the ice famine, and ploughed it into building an icehouse – this time made of brick – in Savannah. It was a grand affair, large enough to hold 200 tons of ice, and all sealed by cavity walls packed with an insulating layer of powdered charcoal.

The Savannah trade began in 1819 and was a great success. The Martinique business had also been revived, and in July of that year Frederic's agent there, Stephen Cabot, reported that he was out of ice and would have to shut up shop. Frederic was in no mood to deal with this problem. His father had just died, and he was mourning the Judge's loss

*The first one lost a mast; the second one sank.

and regretting him dying while his beloved family's fortunes were still so precarious. So Cabot set in action a wild plan. He would capture an iceberg!

In the event Cabot, another scion of a wealthy Boston family, had someone else do it. He contracted a Captain Hadlock, a whaler from Maine, to sail north in the aptly named brig *Retrieve*, and get him a chunk of the Arctic. Even in August when this madcap adventure began, icebergs were common off the coast of Labrador, 'calved' from freshwater glaciers of Greenland and elsewhere. Hadlock found a suitable iceberg in September and sent men ashore (if the iceberg has a shore) to hack off chunks, which floated free of the main mass. All was going well; the *Retrieve*'s hold was nearly full. Then the iceberg rolled over. The ice hackers had upset its centre of gravity. The tumble of ice damaged the *Retrieve*'s hull, but the stakes were high enough for Hadlock to make his way to Martinique – the crew pumping out water all the way. Soon Caribbean customers were enjoying a slice of Arctic ice. The Martinique business was saved, although Cabot appears to have spent all the profits and more on a lavish expat lifestyle.

Frederic, as usual, had more pressing concerns. A rival named Richard Salmon had set up an ice business in New Orleans in 1818, and Frederic was concerned – terrified – that he would lose the biggest slice of the action. For once fate favoured the Tudors – Salmon went bankrupt and died of yellow fever. But Frederic needed investment to move in and take over.

With 15 years of experience behind him, business partners were becoming easier to find. Frederic even accepted money from William and James, the two buffoons who cost him so dearly in that first season in Martinique. With a complex constellation of investors and deals in place, Harry Tudor was sent to New Orleans in the autumn of 1820 to get the icehouse in place.

By the summer of 1821 the Louisiana ice business was bringing in $1,200 a month. But by now the cumulative anxiety of the long hard years had caught up with Frederic. He suffered a physical and nervous collapse and was under the care of his older sister, Emma. Her husband, Robert Gardiner, became a caretaker manager for the Tudor ice business.

Frederic took a working holiday in Havana, sorting out local difficulties – the icehouse managers kept dying. He returned to Boston in 1823 to find that Gardiner had husbanded resources expertly – with a little injection of his own capital – and all of the Tudor operations were in profit, most of the debt was repaid and the business was ripe for expansion. What a difference a year makes! And having someone else in charge …

❅

By the mid-1820s, the Tudors were the biggest players among a growing pool of ice merchants. The competition began at home. No one owned the ice freezing on the rivers and lakes. Whoever got there first would get the lot. Until now ice harvesting was an unrefined process. The ice was simply smashed up into rough chunks and loaded onto wagons for direct transport to the wharves. One trick was to 'sink the pond', which meant drilling holes in the ice as it froze, so water from underneath welled up and flooded the top, thus creating a thicker slab.

For many years, the Tudors had collected ice from their pond at Rockwood and bought further supplies from other local sources. By 1825, one of these suppliers, Nathaniel Wyeth, had developed a system for cutting regular blocks of ice. His contraption was a horse-drawn cutting tool, a cross between a plough and a saw. The ice cutter was pulled by horses, each shod with spiked shoes so that it did not slip. The blade of Wyeth's cutter etched out deep grooves on the ice, marking out blocks of equal size. The cutter was re-run through all these grooves until they were deep enough for the icemen to cut each block free by hand. To do this they used an array of oversized carpentry tools, such as long saws and giant chisels. Once split, blocks were captured with hook lines and poles, and floated to the bank. There they were seized with a pair of immense tongs and hauled ashore.

Wyeth's system was very efficient for harvesting the ice, and it also made it possible to pack more ice into holds and storehouses – the regular blocks fitted together more neatly than the smashed-up pieces that had come before. Such efficiencies would prove invaluable as the competition for ice hotted up in the unseasonably warm winter of 1827–1828.

The odd crop of thin ice, a few inches deep by all accounts, was collected for sale in Boston. However, the long-distance ice trade needed slabs measured in feet not inches, and the demand for ice was rising. Come summer everyone would want it from New York to New Orleans. But the thick ice eluded the deep ponds of Massachusetts. A frantic search for ice began.

Wyeth's men found a large source of thick ice at a place called Swain's Pond. They arrived by night so as not to alert rival teams. However, this incognito approach was well and truly blown in the early hours, when they were forced to use gunpowder to blast away boulders blocking the track for the wagons. As dawn rose, the scene was like a battlefield. Wagons were struggling through the increasingly muddy lane weighed down with ice, while workmen injured by flying rock fragments from the blasting were being treated in a nearby farmhouse. Frederic gave the farmer 25¢ by way of thanks. He paid $10 to another landowner for the right to open up a second front, clearing the trees from the shoreline to make it easier to reach the pond. Another harvester arrived and paid $2.50 for Wyeth's team to cut some ice for him. All in all, the ice cut from Swain Pond cost Frederic a total of $7.75, but that was just to get it on land. The teamsters who were there to transport the ice to the wharf were finding it almost impossible to shift, as their wagons sank in the mud. In response they dug their heels in, frequently demanding higher fees as the iced loads melted away. Sailcloth called for from Boston was draped across trees to keep the ice in the shade as each wagon edged its way to the main road. In the end most of the ice had to be abandoned.

Frederic concluded that perhaps it was time 'to change latitude', meaning move to harvesting operations north, to places like the Kennebec River in Maine, close to where Richard Gardiner lived.

Frederic's prediction would be for others to realise. Wyeth went off to forge a new life – and dice with death – at the other end of the Oregon Trail (and by some accounts to get away from his wife). Frederic, forty-something, newly married and with a handsome though not-yet enormous income, settled into a life of comfort. With his reputation restored and future assured, his interests turned to property deals and a new venture future trading coffee. It is no surprise that the same themes appear in later life. At one point Frederic had options on 15 per cent of the entire North American coffee supply and was looking at a loss of $210,000 – the equivalent of $5.5 million today. However, the story of the Tudor Ice King was already becoming a legend, and that Tudor legend is what brought ice to India.

❋

Frederic Tudor knew there was a market for ice in India, but even he was not bold enough to give it a try. In 1833, Samuel Austin, one of the next generation of Boston merchants, came to Frederic with an idea that would change all that. Austin's ships made regular runs to Calcutta, but were often largely empty on the way out. Boston ships had a perennial problem with finding ballast. Many crews resorted to dragging the harbour in search of big stones that they could use to weigh down their vessels on the empty outward runs to far-flung exotic locations.

Austin's idea was to use ice as ballast. Unlike rocks – India had its fair share – the ice that survived the journey, which involved crossing the equator, rounding the Cape of Good Hope, then going back across the equator, could be sold to the British colonials.

In 1833, an East Indiaman, the *Tuscany*, one of the prides of the Boston fleet, was loaded with 180 tons of Tudor ice, but Frederic only took a third share in the enterprise. The *Tuscany* left in early May with strict orders not to open the hold to check the ice. In September it arrived at the mouth of the Hooghly River, where it would need to take on a pilot to navigate upstream to Calcutta. It took eight days for the ship to creep up the river, with some reports saying that it was seen to rise out of the water by several inches as its cargo melted inside.

The people of Calcutta waited for the arrival of the ice with growing anxiety and excitement. Ice was already available to buy in the city, made by the age-old evaporative cooling techniques, which caused a glaze of frost to form on the sides of earthenware jars. However, this 'hooghly slush' was no match for the crystal-clear slabs from New England. As the *Tuscany* reached the quay, the authorities had already waived customs duty and allowed for the ice to be unloaded at night – normally something that was against the rules. And so enamoured were the Anglo-Indians with the frozen cargo (there was still quite a bit left) that they arranged for an icehouse to be built for the next supply run. Calcutta was known as the City of Palaces, and the icehouse was to be an edifice fit to suit its surroundings.

The ice trade was going global with shipments heading for China, New Zealand, Australia and Brazil. By 1856, 146,000 tons of American ice were being transported to India alone, and new markets had opened up in Europe. Carlos Gatti, a Swiss-Italian businessmen living in London, was an early entrant. He combined the Swiss tradition of chocolate making with the Italian one for ice creams. Gatti's cafe was the first to sell ice cream to the British public. Initially it was made from ice harvested from the Regent's

Canal. Remember, this ice was not consumed with the dessert, merely used as the coolant to make it. One can only guess at what material was mixed into the canal water. Gatti's desserts caused something of a sensation at the Great Exhibition of 1851, and as his business boomed he built an ice house on the canal near King's Cross. Today, this building is the London Canal Museum.*

Back in New England, vast icehouses were being built beside ponds and lakes to store winter ice the whole year round. Slabs were shaped, smoothed and cleaned by machines before conveyers and cranes – some designed by Nathaniel Wyeth safely returned from the frontier territories – lifted the ice to the roofline. Just as in the first Tudor icehouse in Havana, ice was loaded from the top down.

Ice was no longer free. Harvesting rights were bought and sold, with the area of each lake divided according to who owned the rights for which portion of bank. Every source of ice was becoming a valuable brand: Spy Pond, Fresh Pond, Jamaica Pond.

In the grand houses of London, ice from Wenham Lake near Salem, Massachusetts was regarded as the only ice to have, being clean, clear and positively healthful. In fact, by the time it got across the Atlantic just about all American ice had become a supply from Wenham. In Norway, Lake Oppegaard near Oslo was simply renamed Wenham Lake in an attempt to get in on the action.

Wenham ice even gets a mention in *The Second Jungle Book*, written by Rudyard Kipling during a stay in Vermont. A story within relates how an adjutant bird ate some of the ice that fell overboard from an American ship, unloading on

*Gatti's ice cream and chocolate empire made him wealthy, but it was not his first love. When an insurance windfall arrived in 1862, he invested the money in Gatti's Music Hall. He sold it five years later to a railway company, which razed it to the ground and built Charing Cross Station on top.

the Hooghly River. The bird describes the mysterious pleasure of the cold and then the terrible feeling of loss when it was gone with the ice reduced to nothing but water:

> *I danced and cried out against the falseness of this world; and the boatmen derided me till they fell down. The chief wonder of the matter, setting aside that marvellous coldness, was that there was nothing at all in my crop when I had finished my lamentings!*

The ice industry was not welcomed by every human either. In the mid-1840s Henry David Thoreau had built a cabin beside tranquil Walden Pond near Concord, New Hampshire. Walden was a place to contemplate, away from the distractions of modern life – you know, all those incessant telegrams. However, modern life followed Thoreau to his bolt-hole. In 1846, the railway reached Concord, and with it came Tudor's men to Walden. Rail was replacing road as the way to transport ice to port. Thoreau described the activities of the ice harvesters on the pond in his seminal work, *Walden*, published in 1856. Largely ignored at the time, this book belongs to the transcendentalist movement, which is an all-American philosophy that questions industrial society and in many respects forms the roots of the modern green movement – and arguably the survivalist movement as well. Thoreau trusted nature not modern society. He thought it absurd that someone like Frederic Tudor should grow fabulously wealthy by collecting ice from a lake, and questioned whether it did him and society any good.

By the 1880s more than five million tons of ice were being consumed in the United States alone – and that figure was set to rise further. Also on the rise were epidemics of typhoid and dysentery, the blame for which was being placed firmly on natural ice. Thoreau was right, the ice industry was not doing us any good. But as we began to distrust natural cold, could we learn to trust a mechanical one instead?

EIGHT

Taking the Heat

Without dependable performance the savings and the convenience of modern refrigeration are lost. The G-E mechanism is entirely automatic and hermetically sealed inside walls of steel.

General Electric Monitor Top advertisement, 1935

It's now time to talk about the refrigerator, the fridge, the cooler, the icebox. Admittedly, it's taken a long time to get here. Firstly, that's because of the many centuries it took to wrangle the scientific principles and the concurrent struggle to master the technology that harnessed them. Secondly, it is because artificial cooling only took off when it became clear that the natural version was no good at the job. Nature furnishes the planet with more than enough ice to meet our needs, so why don't we use that? Why did Frederic Tudor's dream die? For die it did, albeit a long, cold death.

In the nineteenth century, it was a universally held belief that natural products were pure and clean, and that went doubly for natural ice. Even after John Snow showed that cholera was spread by London's water in 1849, and after Louis Pasteur developed the germ theory in the 1860s, which revealed that invisible microorganisms were responsible for disease, people clung to the belief that the freezing process somehow neutralised or expelled any impurities.

The reality was very different, as Frederic Tudor's early customers would have seen. A chip or two of New England ice certainly chilled a drink, but as it was consumed, you would have become increasingly aware of a dark sediment gathering at the bottom of the glass. The New England lakes are indeed famed for their clear blue waters. They produce crystal-clear ice, with a tinge of green. The green is from plant-derived chemicals washed in from soil, plus a dose of algae living in the water. In this way, lake ice superseded river ice in two respects. The slow current makes most river water a little turbid, so the finer silts are trapped in the ice when it freezes. There is also a lot more air mixed into the moving water, which manifests itself as tiny bubbles, and that makes the ice opaque, and less 'natural'.

'Sinking the pond' by drilling holes in the ice so that water welled up from underneath increased the ice yield but reduced the quality of the ice. The fresh surge of water trapped whatever sediments had gathered on the surface of the ice – leaves, dust, an iceman's footprint. No matter how scrupulous the cleaning of each cut slab, that stuff was inside until the ice melted.

It was true that ice from high latitudes, where the water froze fast and deep, did not suffer these problems. But the likes of Wenham Lake and the Kennebec River could not supply the whole world, despite what it said on the bill of sale.

Cheaper ice was cut from rivers upstream from the big American cities. As the demand for ice grew, harvesters became less choosy and began collecting ice closer to town, frequenting waters that mixed with sewage and industrial outflows.

In the 1880s ice cut by Philadelphia's Knickerbocker Ice Company from the Schuylkill River was a dark green. The river was the drain for the city's slaughterhouses and textile factories, and it was so polluted that the banks were covered in dead fish. No one wanted green ice. Knickerbocker was forced to import ice from Maine – branded as 'eastern ice' – putting the price up and opening the door to alternatives.

Alternatives were available. For forty years, the Tudor Ice Company had been a monopoly provider to the big coastal cities of India, Bombay (Mumbai), Madras (Chennai) and Calcutta (Kolkata). These cities had the world's most ornate ice houses, looking more like royal palaces than the whitewashed sheds of New England. However, the reign of the Ice King was short. In 1878, a steam-powered artificial ice factory was built in Kolkata. Within five years the American ice ships had stopped coming.

The ice palace in Madras, however, found a new calling. In the 1890s it was briefly the home of Swami Vivekananda, a prominent Hindu holy man, and it has subsequently become an institute (Vivekanandar Illam) devoted to his life's work. Vivekananda was one of the first to promote Hinduism in the United States and became something of a wandering ambassador delighting crowds with his impassioned speeches. His most famous talk was at the first World's Parliament of Religions in 1893 during Chicago's World Columbian Exposition, which caused something of a stir. Elsewhere, the Exposition was showing off artificial refrigeration. That caused quite a stir as well, but for all the wrong reasons, as we shall see later.

Artificial ice was a viable business in the Indian market, where ice was always a luxury for the wealthy colonials, who could afford it whatever its provenance. Additionally, in the American South, where the most profitable ice markets of New Orleans and Charleston had been cut off from the Yankee ice supply during the Civil War, artificial ice was on the rise – especially inland where natural ice seldom reached.

By contrast, in Philadelphia and New York, natural ice was always able to undercut the artificial in price. There remained a market among the poorest in society, who would be willing to take the risk and spend their small change on a block of cool during the summer.

Eventually, however, the link between natural ice and diseases, typhoid chief among them, would prove too great. Human faecal matter in water had been linked to typhoid in the 1880s, but the science was still sketchy. The death knell for natural ice was struck by a typhoid outbreak in, of all places, a hospital. In 1902, the staff at a mental hospital in Ogdenburg, New York, had the ice from the nearby

St Lawrence River harvested for use the following summer. The result was a typhoid epidemic that killed several patients and staff. The ensuing investigation showed that the stretch of river where the ice was cut was fed by the hospital's own sewer.

Artificial ice companies were already explaining the dangers. Promotional material from 1899 says: 'Natural ice contains, always, a certain amount of impurities, gaseous, and solid, no matter where it's obtained from.' A later warning was more blunt: 'Don't take the iceman's word for anything.'

Icemen were an untrustworthy bunch in general. The title of Eugene O'Neill's play *The Iceman Cometh* is not just an allusion to a dirty joke; it is also a metaphor for the empty promise of empty dreams that the characters in the play must confront.

Icemen are a thing of the past now, but at the time they occupied a position in the popular imagination similar to that of milkmen (themselves a dwindling breed). In the 1960s and 1970s it was often joked that milkmen were cuckolding husbands while they were out at work. The icemen of the late nineteenth century had the same opportunity, and husbands were doubly suspicious of them because icemen were big units, hunks, tough guys. They had to be to manhandle the heavy ice, carrying it up and down stairs, all the while wearing thick woollen clothing to protect the skin from the great cold (coalmen were big too, but presumably too filthy for any dirty stuff). Joe Louis, the heavyweight champ from 1937 to 1949, built his muscles as an iceman in Detroit before punching his way to the big time.

Icemen were trouble for their employers as well. They could demand high wages for their services and could easily cream off profits, because stolen stock would simply have melted away. For ice companies, the deliverymen cost more than buying the ice itself.

A job as an iceman was precarious enough with so many suspicions following them around, but it was also seasonal.

In the 1932 Marx Brother's film *Horse Feathers*, Chico and Harpo play icemen. However, they need to supplement their income, and in keeping with the low opinion of icemen, Chico becomes a smuggler, while Harpo is the town's dog-catcher.

To begin with at least there was no natural ice to sell in winter – although the demand remained. From the 1830s onwards, American households became used to keeping dairy products and other foods fresh inside an icebox. The great benefit was that it kept food safe to eat for longer and it also remained tasting fresh.

Back then iceboxes were frequently called refrigerators, but we would not recognise them as such. The concept was simple enough. A block of ice was placed in a container at the top of the device. It gradually melted, and water trickled into a tray at the bottom. However, the ice did cool the air inside the large lower chamber – as long as the door was kept shut. The most apparent functional difference with today's fridges is that an icebox could never freeze anything.

Iceboxes were seldom very effective. The main focus was on how they looked. Iceboxes were generally too large to fit in a small nineteenth-century kitchen and were therefore part of the furniture in the family parlour or dining room. Some designs even had the box sunk into the floor, only for it to rise up when commanded using weights and pulleys. Icebox insulation was often ineffective, and ice did not last long in summer – hence the frequent question: 'Has the iceman come yet?' The reader can imagine the many and varied possible answers, but if the answer were simply 'no' then that meant that the family's butter, milk, fish and meat were probably all going bad.

Artificial ice companies were more expensive but more reliable with their deliveries – and kept them coming all year round. In addition ice made by artificial means was crystal clear and looked more natural than the natural stuff. The water was boiled before being frozen, which sterilised the bugs and removed the air. For the icebox owner, artificial ice was cleaner and more reliable than natural ice (ironically, the marketing men would say precisely the same thing to convince the public to replace iceboxes with refrigerators).

However, there was still a problem with artificial refrigeration systems. They tended to explode.

Until that issue was resolved, refrigeration would remain a service delivered to the house as ice. The icemen would have to keep on coming, with ice of an unknown, unproven source for an icebox of limited effectiveness. We really needed fridges – or 'self-feeding ice safes' as we might have known them then. To find out how we got them in the end, we have to rewind the story, back once more to 1805. We are in the United States again, though not in Thomas Moore's Maryland or Frederic Tudor's Massachusetts, but in Philadelphia, the home of the prolific inventor Oliver Evans.

❋

By 1805, Oliver Evans was already an esteemed engineer with many design plaudits to his name. He had almost single-handedly industrialised the production of flour in the eastern United States with an automated milling system. The American colonies before the revolution had struggled to produce decent flour. The hard grains that grew there were difficult to process by traditional milling methods, and required a large workforce to grind and sift them. The workers inside the mill were part of the problem, too, frequently contaminating the flour with dirt and mixing up the process. The result was often a coarse, gritty flour that could not compare with supplies coming over from Europe. Evans, a young wheelwright from Delaware, devised an ingenious constellation of machines that did it all automatically, under the motive power of a waterwheel. Despite many years of resistance from millers, one of whom described the contraption as a 'set of rattletraps', the value of Evans's designs was eventually recognised by the post-revolutionary federal government. By the late 1790s Evans's mills were being built across the Eastern Seaboard to help feed the young nation.

By now Evans had moved to Philadelphia, the seat of the US government at the time, and turned his attention to steam power as an alternative to water. He was not alone in dabbling in this technology, and duplicated many of the advances made by Richard Trevithick, James Watt and others working in Europe and North America. Evans's unique contribution to the field was the Oruktor Amphibilos – less obscurely described as the 'amphibious digger'.

The concept was a steam-powered dredger for use in improving harbours. There are few accounts of what actually happened to this monstrous craft. It was about 10 metres, (32 ft) long and weighed 17 tons. The main body was a flat-bottomed boat fitted with dredging equipment and powered by a rotary paddle at the stern. However, it was amphibious, and the entire thing sat on four wheels that were also connected to the steam engine filling the bow. Whether the craft made it to the water is unclear, although we do know that it was so heavy that on at least one occasion its wheels collapsed. Nevertheless, the Oruktor Amphibilos is regarded as the first self-propelled vehicle to have graced the roads of North America – however short its journey. If it did indeed make the Schuylkill River (no one knows whether it did for sure) then it was also the second American steamboat.

Evans's experiments with steam engines made him an expert in the heating, compression and expansion of gases. Nevertheless, his expertise was only in the context of the time. These were the days before Carnot, Mayer and Joule, when the workings of a steam engine were something of mystery. The focus for Evans and his like was to tame its natural force rather than understand it. In 1805 he published his knowledge to date in the *Steam Engineer's Guide*. Although the version that went to press was still rather incomplete, the book was nevertheless popular. It made a good fist of introducing high-pressure engine technology to a wider audience, but the theories and principles it set out were all wide of the mark and of no particular use to future steam engineers.

More interestingly, in the back of the book Evans ponders the power of steam engines to create cold. He would have been aware of the evaporative cooling experiments pioneered by William Cullen, not least because they had been repeated soon after by Benjamin Franklin, the archetype of American scientists. Remember, Cullen used a vacuum pump to make a liquid evaporate very quickly, creating cold as it did so. Evans described how the mechanisms of a steam engine might be used to make this a continuous process, where the evaporated gas was compressed back into a liquid by a steam-powered piston, then allowed to evaporate again and again. Evans suggested ether would be a suitable substance to use, although he never sought to build the device he proposed. He had nevertheless described, in a nutshell, a vapour-compression cycle, the very system used in refrigerators ever since.

Although no one understood it back then, this is how it works. The fluid used is called the refrigerant. Evans thought of ether, but many other substances were tried over the following 125 years that it took to develop reliable refrigerators. The refrigerant is contained in a closed loop, effectively a single pipe that goes around in a circle, although it is orientated more intelligently than that to fit inside your fridge. But you get the picture. The pipe connects to four main components: the compressor, the expansion valve and two sets of heat exchangers (which are basically just more pipes).

The vapour-compression cycle is just that, a cycle, so it has no beginning or end, but let's start at the compressor. The refrigerant enters the compressor as a gas, where it is squeezed into a smaller volume. The piston's motion is transferred to the gas particles, making them move faster, and that makes the gas hotter. The hot gas leaves the compressor and enters the first heat exchanger, often called the condenser. The purpose of this component is for the heat from the hot gas to escape into the surroundings, leaving the system forever. To this end the heat exchanger is designed to maximise its contact with the outside world, and that's the long, twisting coil of narrow pipes covering most of the back of the refrigerator. Don't risk an injury by hauling the fridge out to take a look, just put your hand around the back – it's warm. That's not because the power unit is working hard and hot, it's the system shedding heat. A modern fridge probably has a small fan underneath which is blowing air over the exterior coils to enhance the cooling effect.

By the time the hot gas reaches the end of the heat exchanger it has given up its heat and condensed into a cool liquid. The temperature has dropped but the pressure has not. The compressor is effectively a pump that pushes the refrigerant around the system, so the liquid refrigerant continues to the next stage, the expansion valve.

The expansion valve is relatively simple. It's a small nozzle through which the high-pressure liquid is squirted. Thinking back to those hard-won gas laws, an increase in gas volume creates a decrease in pressure. But because the gas is not behaving 'ideally', its particles have to work to move away from each other and that slows them down, making the gas colder. This is the so-called Joule-Thomson Effect, as revealed by that eminent pair in 1852.

The same thing is happening in the expansion valve only in a much more efficient way. The liquid expands and evaporates inside a chamber on the other side of the valve. Calling on historical figures once more, Joseph Black's latent heat is at play here. The state change from liquid to gas requires an extra injection of heat energy over and above a basic temperature change. That means that for a refrigerant to be effective it needs to have a high latent heat, to maximise the amount of heat it steals as it evaporates.

So the liquid refrigerant has become a very cold gas, and on it goes to the second heat exchanger, which is also called the refrigeration coil. You will not find this one on the outside of the refrigerator. It is hidden away inside behind the back panel of the food compartment. This compartment acts as the heat source with thermal energy shifting to the cold gas in the coil (the heat sink). The source and sink reach a thermal equilibrium of sorts with the gas warming up and the compartment cooling down. The effect is further enhanced because freshly cooled gas is constantly entering the coil so the compartment is always giving away its heat. A modern fridge has a second fan at the top of the compartment that draws the air down over the cold pipes of the exchanger and back in again at the bottom. This only happens when the door is shut and it also serves to remove damp air that has been let in from the outside.

Readers over 50 will be well aware that in older fridges, made before fans were added to refrigerators, moisture in the air would gradually build up inside the compartments and on the internal coil and require defrosting. Modern frost-free fridges have a small heating system linked to the refrigerator coil. This turns on every few hours and melts any ice that has built up. That might interrupt the chilling process for a short while, but it boosts overall efficiency in the long run.

So, the refrigerant leaves the chilling coils and has arrived back at the compressor. Off it goes round again.

Such a device can be used as an ice-making machine, or it can reduce the temperature of a space and be what we understand today as a refrigerator. The market for ice was already established, so the former application made more sense to the pioneers of refrigeration, but the mechanism for both is the same.

❋

In August 1835, the following patent was issued in London:

> *Now know ye, that in compliance with the said proviso, I, the said Jacob Perkins, do hereby declare the nature of my invention, and the manner in which the same is to be carried into effect, are fully described and ascertained in and by the following description thereof, reference being had to the drawings hereunto annexed, and to the figures and letters marked thereon (that is to say): It is well known that by evaporating volatile fluids from the surfaces of vessels containing other fluids, the caloric of the latter fluids is reduced, and cools down the temperature indicated by a thermometer immersed therein; but in so evaporating volatile fluids the same is lost, and hence the application of this process to produce a considerable degree of cooling effect to fluids requiring that process on a large scale, a large cost is necessary, and hence renders this means of cooling of little practical value.*
>
> *Now the object of my invention is so to use a volatile fluid that the same (having been evaporated by the heat or caloric contained in the fluid about to be reduced in temperature) shall be condensed and come again into the vessel to be again evaporated and carry off further quantities of caloric.*

Fields of endeavour in science and engineering are often given a 'father', a towering figure who is credited with their foundation. We could say that William Cullen was the great-grandfather of refrigeration, and that Oliver Evans was the grandfather. In that case Jacob Perkins would be its father, although not a very good one.

Perkins built an ice-making machine in London in 1834. He got the idea straight from Evans, who had died 15 years before. The pair had worked on steam engines in Philadelphia, and no doubt discussed it. Perkins was a printer and a very good one at that. In 1819 he went to London in the hope of winning the contract for designing banknotes that could not be forged – counterfeited British currency was a global problem. Perkins's designs were exquisitely intricate, but his American heritage meant he was barred from working in a role so crucial to the security of the British Empire.

To get around that obstacle Perkins formed a partnership with two Englishmen, and the venture was a great success, with Perkins producing currency and postage stamps for many nations.

Perkins was also interested in steam power. He designed a steam gun that fired 1,000 rounds per minute using the force of high-pressure steam. It was reportedly rejected by the Duke of Wellington for being 'too destructive'.

The ice machine was the opposite; it was too inefficient and could only make small amounts of ice – which no one in England wanted at that time, anyway.

Despite a successful business Perkins was a less than successful businessman and had no money to take the ice machine further. This job was taken on by a couple of 'uncles' of refrigeration. Alexander Twining, a professor at Yale University, improved the compressor and set up an ice plant in Cleveland in the 1850s. It produced half a ton of ice a day, but nobody in Cleveland wanted it either – natural ice was much cheaper.

A Florida doctor, John Gorrie, had stumbled upon another technique for making ice while investigating ways of keeping feverish patients cool at his sultry Apalachicola hospital. He built a compression system to chill water. The intention was to use this cold water to cool the air of a room, but it worked so well that the water froze. Gorrie quit medicine, got the first US patent for refrigeration and entered the ice business. He set up ice machines in New Orleans and Cincinnati, but lost his money.

The problem was that artificial ice made with huge, haphazard machines was just too expensive compared to the lake ice delivered from the north. However, there was one place where even the Ice King of Boston could not dominate: Australia.

Natural ice and snow were rare on the mainland. Only Tasmania got cold enough weather with any regularity, but it was nothing like the ice powerhouses of Scandinavia or New England. On a typical voyage from Boston to India, a third of the ice would melt in the hold. Shipments did go further and reach Australia, but with most people living in the south-east corner of the continent, by the time the Boston ice ships got around there the frozen cargo was much depleted.

In the 1850s, a Scottish journalist, James Harrison, had moved to Australia and was running a newspaper in Geelong, a town west along the coast from Melbourne. On our family tree, Harrison would be the stepfather of refrigeration. He took on the problem and made something of it. Ether was part of the tool-kit of a newspaper man back then. It was used to clean the ink from printing plates after each edition had hit the news-stand. Harrison's interest in refrigeration, for which there was an obvious demand, is reported to have been inspired by the way the evaporating ether made the metal type feel icy cold.

Harrison took on the ether-compression designs of Perkins, Twining and Gorrie and made them work – his best machines were manufactured in England and sent out in sections for assembly in Australia. Harrison's main breakthrough was a compressor driven by an immense five-metre flywheel. That made it powerful enough to produce a significant temperature drop when the ether expanded.

Harrison's first machine, built in 1851, was used to make ice. However, his next machine was the world's first functioning refrigerator. Fittingly for this Australian triumph, this historic refrigerator was used to chill beer.

An ether machine like Harrison's was immense. It would struggle to fit inside a two-storey house. Harrison made a success of his designs by fitting them into cargo ships, where they kept Australian meats and other produce fresh for the long voyages to markets overseas. However, a smaller refrigerator would need a different refrigerant.

Ammonia was top of the list. Ammonia is not quite as volatile as ether but has a far greater latent heat. To produce the same temperature drop, a compressor using ether has to be roughly 17 times bigger than one using ammonia.

Both Harrison and Twining had taken patents on using ammonia as a refrigerant but neither put them into action. The first refrigerator to use ammonia was invented in 1859 by Frenchman Ferdinand Carré. His design took a Gallic swerve on the basic compression model (his brother Edmond had actually invented it in 1850, but his one used sulphuric acid, which was even trickier to handle than ammonia). The refrigerant in Carré's device was actually aqua ammonia, a fancy name for ammonia dissolved in water. A boiler gently heated the aqua ammonia, making the more volatile ammonia evaporate off. The pipe with this warm gas then passed through a tank of cold water, which condensed it into liquid ammonia. The liquid was then evaporated using an expansion valve, and the resulting cold gas entered the refrigeration coil, chilling water or whatever else. Finally the cold gas re-entered the boiler where it promptly dissolved again. Such a system was called an absorption refrigerator. It had one big advantage over using ammonia in a compression system. All that water made it very unlikely that the ammonia would explode. Carré's absorption system was the only viable system well into the 1870s. He built a refrigerated ship, the *Paraguay*, in 1876, which was able to keep South American beef frozen solid for the entire voyage across to Europe.

Both ether and ammonia are flammable and the search for an alternative refrigerant continued. Thermodynamics was still in its infancy, so developers simply tried their luck with likely materials. Consequently progress was slow.

In 1874, a Swiss professor of physics, Raoul Pictet, built a compression refrigerator that used sulphur dioxide. This gas does not burn, but that is about its only advantage. The latent heat was low so it needed big machines working hard. It also smells awful. A brief whiff smells like mould, a larger one catches the throat and makes you gag. The smell is perhaps a godsend because without its warning a leaking sulphur dioxide fridge would kill you – the gas is also pretty toxic.

Another alternative tried was carbon dioxide. This gas was cheap, nonflammable and only killed you if you took a big suck of it. It even doused fires. However, as when its chemical cousin, sulphur dioxide, was used, machines with carbon dioxide ran at enormous pressures, and frequently developed leaks.

It took a German to sort all this out. In 1873, Carl Linde, a professor extraordinarius specialising in thermodynamics at the Munich Polytechnic, developed a compression refrigerator system that was dependable and efficient and could be made small enough to fit in a normal room – perhaps even one day inside a home. At first the refrigerant he used was dimethyl ether, but he switched to ammonia as he began taking orders for ice machines from lager breweries across Europe. Linde set up Linde's Eismaschinen AG, or Linde's Ice Machine Co. It became a market leader in ice machines mostly because it licensed the technology to manufacturers across Europe and the United States.

Linde did not enjoy being a business magnate and was more interested in the science. He developed new systems for exploring the depths of cold and its effects on matter. His company left others to exploit the refrigerator market and instead it used the Linde technology to supply the world with the refrigerants it needed and many other industrial chemicals, such as purified air gases. This is one of many examples of how refrigeration impacts our lives outside the kitchen as well as in it.

With Linde's design and supply of refrigerant, the stage was set for refrigeration to take over the world. It would take another 40 years before domestic refrigerators were a viable option for the ordinary home. However, the technology was embraced by the food industry and ice companies, which saw all the benefits and none of the risks. Nevertheless, consumers were more cautious about the prospect of making ice by machine. How did it work (few people were able to explain it), and was it safe?

In 1893, the World's Columbian Exposition was chosen as a good place to show America the advantages of the new ammonia compression ice machines. But this did not go to plan.

❅

When Captain James Fitzpatrick of the Chicago Fire Department heard that there was another fire at the World's Columbian Exposition, he probably already had a good idea where it was. The ice-making plant had caught fire twice before. The plant was not really part of the expo's pavilions, which were showing off the latest technologies from all over the world. However, at the behest of the burgeoning refrigeration industry, a public display of the technology was set up inside the plant making ice for the visitors.

The plant was certainly the centre of attention by the time the firefighters arrived. Yet again the chimney had caught fire, or rather the wooden tower that had been built around the smokestack, which had been deemed too ugly to grace the fair's temporary skyline. The tower design had called for a metal grill to cap the smokestack inside and contain any sparks and embers that erupted from the coal furnace inside. Only the grill had never been fitted – and the thing was on fire again.

Fitzpatrick and his men used ladders to reach a balcony a few metres below where the flames could be seen. The firefighters assumed that it was just the stack that was on fire, as it had been the last two times, and all that was required was a bit of well-aimed water to douse it. However, this time the fire was spreading inside the building as well. Moments after the firefighters reached the balcony, the building exploded. Men were trapped by the flames. Many jumped, others attempted to climb to safety. Most fell or burned, or did both. In total 17 people died, including the captain, and 19 more were critically injured.

No one is quite sure what caused the explosion, but it is more than likely to have been high-pressure ammonia mixing with the air inside until it reached a critical concentration – and then boom! The fire did a great deal of damage to the reputation of refrigerators, and not without good reason. The explosive scenario was a little freakish, but refrigeration plants would continue to be plagued by fires from leaking pipes. When electrically powered refrigerators became available in the 1910s, many people still preferred to stick with ice. They knew where they were with ice.

The saving grace of the artificial refrigeration industry, in America at least, was the fact that fire was a problem for the natural ice industry too. By the 1890s, there was a new ice king on the throne, Charles Morse from Maine, whose Kennebec River ice was the market leader. Morse used his position to buy up just about all of the ice companies from Maine to New York. By 1900, his American Ice Company was a virtual monopoly worth $60 million. That valuation was based on Morse's ability to double the price of ice overnight, which he was prone to do.

Morse was no friend of the common man, but he had a reputation for getting close to New York's elite. What few rivals of the American Ice Co that remained soon found that their icehouses and wharves were demolished or condemned by the city for being hazardous. An investigation revealed that the mayor and many of his officials and several New York judges held large amounts of American Ice stock. They were in the company's pocket. Morse was called to Washington to explain himself. He declined the invitation and resigned from the company with a few million stashed away. The second and final ice king would eventually end up in jail, just like the first.

American Ice stayed intact, but 'open winters' with no ice and a continued decline in water quality put pressure on the profits. The beginning of the end came in 1910 up on the Kennebec River. A train passing through Iceboro (a town built on the ice industry) threw a spark that set alight an American Ice Company icehouse on the bank. Despite being full of frozen water, the giant icehouses, whitewashed to reflect away the sun's rays were highly flammable due to all the sawdust used as insulation inside. The winds spread the first fire to Iceboro's many other ice stores. About 40,000 tons of ice went, along with a couple of ships that got stuck in the shallows. The natural ice industry would never recover.

❄

Artificial ice was already taking over anyway, with perfect slabs of perfect ice sliding out from freezing pools. There was a boom in these plants when the cheap ammonia refrigerant came on the market in the 1880s. Many ran into the 1970s and beyond. The plants produced cold in the same way as any fridge, except that the refrigeration coil snaked through a deep tank. This tank was a little more than a metre deep and about a quarter of the size of an Olympic swimming pool. The big difference was that it was filled with brine. The brine was super salty so it stayed liquid even when the refrigeration system took it down to about -10°C. Freshwater was poured into an array of steel moulds, which were carefully lowered into the brine. The entire tank was then covered over to keep the cold in. It would take a couple of days for all the moulds to freeze, at which point they were lifted out – they were arrayed in long lines that could be hoisted up by a pulley system – and moved to a smaller dunk tank filled with fresh, heated water. A short spell in there freed the blocks of ice from the metal moulds so they could be tipped out on to a gently sloping wooden floor. The blocks were then manhandled into a storehouse where they were sawn into smaller blocks that would fit an icebox. The icemen then left to do their work.

In North America, the icebox remained the dominant form of domestic refrigeration until well into the 1920s. In 1914, British travel writer Winifred James commented, 'Whoever heard of an American without an icebox? It is the country's emblem. It asserts his nationality as conclusively as the Stars and Stripes float from his roof-tree, besides being much more useful in keeping his butter cool.' To non-Americans, the icebox remained something of a novelty. The old food-preservation techniques were still in use, and cool outdoor larders and meat safes in the cellar remained the norm until the prosperity of the 1950s. The British were especially recalcitrant. In 1965 only a third of UK households had a fridge, even fewer than the French.

The invention of the domestic refrigerator is credited to a French monk, Abbé Marcel Audiffren. He had been tinkering with a machine that could keep wine cool, and there are accounts of working models using sulphur dioxide as early as 1903. However, the rights to his electrically powered unit, which was designed specifically for use in a home not a monastery, were acquired by General Electric in 1911. It was put on sale for $1,000 dollars, twice the price of an automobile, and like cars, only within the range of the wealthy few.

General Electric were not the only people offering domestic refrigerators. In 1916 Kelvinator machines took an early lead, despite their models requiring customers to cut a hole through the kitchen floor so pipes could connect the body of the fridge to a separate condenser that was installed in the cellar. Kelvinator were followed by Frigidaire in 1919. Both these companies were sponsored by car companies: Buick was behind Kelvinator and General Motors had an interest in Frigidaire. Frigidaire technology was repurposed into air conditioning for GM cars from the 1950s.

One of the most iconic early refrigerators was the G-E Monitor Top, which had the cylindrical compressor and condensor sticking out of the top. The sturdy machine was named after a Civil War ironclad battleship, but the design also made it look a little like a boxy faceless robot parked in the kitchen. The heft of the Monitor Top went a long way to reassuring the public that fridges were safe. As the advert said, the Monitor Top was so dependable that they 'sealed it in steel'. A cynical subtext would read: 'It's completely safe – its surrounded by a metal shield.'

Advances in manufacturing technology had indeed made the refrigerator much safer. It did not leak like the industrial behemoths of the previous century, and that meant there was no smell and no fire. The electric motor was powerful enough to handle the three litres or so of refrigerant inside. The compressor it drove would spin around several billion times in its lifetime. The motor was also affordable to run, although a struggling Ohio ice company pointed out that running a domestic refrigerator used the same amount of electricity as 29 vacuum cleaners. This was slightly disingenuous. The fridge is much more efficient but it is on most of the time, while a cleaner is mostly off (and, of course, in those early days one of the reasons for selling electrical appliances to domestic consumers was to encourage them to become electricity consumers as well).

The motor and its compressor were the only moving parts and they had to be quiet. This was achieved then as now by mounting the whole thing in springs bathed in oil. That absorbed the wobbles and knocks that would otherwise echo through the machine.

Monitor Tops are antiques today, but many still work. Looking at one, it is a little surprising to know that not much has changed since they were the state of the art. The compressor has been stashed away inside, and the condensing coil is now on the back. Older readers may remember that a fridge had an icebox up at the top, which served as the small freezer compartment. This entire thing was surrounded by the refrigeration coil pressed from sheets of aluminium. This was the bit that got cold, and the chilled air sank down to the trays at the bottom – the warmest section for the fruit and veg. In a modern fridge-freezer, this section is hidden away generally behind the panels, most in contact with the lower freezer section. American-style models, with their double doors, have separate systems for each side. Europeans arrived later to the refrigeration market and their refrigerators have always remained considerably smaller – more or less half the volume of those of Americans.

In early fridges the expansion valve connected to the refrigeration coil was exactly that, a twist valve like the one on top of a gas bottle. This could freeze up easily so was replaced with a coil of capillary tube, which did the same thing but was easier to keep ice-free by using some heat from the condensor.

The compression system must not run all the time. Everything in the fridge would freeze. So a thermostat is used to keep it at the desired temperature. This tells the compressor when it's time to stop and when to start up again. A modern fridge is controlled by a microprocessor, basically a simple computer. It gets information about the temperature of the compartments from a thermistor, an electronic component that changes in conductivity – how well it conducts electricity – with temperature. Basically it is an electronic thermometer hard-wired to the fridge's brain.

Until recently, the fridge thermostat had a mechanical design not too dissimilar to a thermometer. A thin, flexible tube filled with a volatile liquid was fitted inside. The liquid expanded and contracted as it warmed up and cooled down. When the compartment was too warm, the liquid expanded into a rubber diaphragm, and that pushed on a switch that got the compressor working. The compressor would run for a set period and stop – and stay stopped until the thermostat's diaphragm began to expand once again.

Of course, the thermostat would be switching the fridge on all the time if the warm air from outside could get in. The body of the fridge is heavily insulated, often with a layer of fluffy fibreglass similar to roof insulation. The weak point is the door. Early fridges were massively built so their heavy doors could be closed very tightly, squeezing the rubber seal airtight. To prove just how strong it was, Frigidaire arranged for a four-ton elephant to stand on one of its models. The door still opened.

A modern fridge would not cope with an acrobatic elephant, but the seal is just as effective. The innocuous concertina of rubber conceals a powerful magnetic strip that actively pulls itself to the metal surround when the door is shut, squeezing the seal closed.

Apart from that nothing much has changed. Then as now, the driving force behind fridge innovation was largely to do with market competition. New gizmos and gadgets are installed inside and out to make the fridge feel better designed, more functional and thoroughly modern. The ability to make ice cubes at home was the refrigerator's 'killer app' of the 1920s, and we have not really moved on much from that.

However, there has been one significant change since the days of the Monitor Top. That is of course the refrigerant. In 1928, the chemist Thomas Midgely – the man whose legacy also includes lead petrol – was commissioned by General Motors to find a less risky refrigerant. His answer was a range of gases called chlorofluorocarbons (CFCs), or to give them their later trade name, Freons. The idea was that CFCs were chemically inert. Each molecule is a chain of carbon atoms studded with chlorines and fluorine atoms. The last two elements produce very strong bonds, and researchers could not find any chemical reaction that would disrupt them – in conditions one would expect to find in the natural world, at least. So CFCs were deemed harmless to humans and to nature but also had an acceptable latent heat and volatility for use as a refrigerant. To start with it cost Du Pont, the eventual owners of the Freon patents, about $50 to make a pound of CFC refrigerant, 2,000 times the price of sulphur dioxide. The company decided to take the hit, and it worked. By the end of the 1930s, CFCs were making a profit. By the 1970s, Freon refrigerants were almost universal among domestic appliances (ammonia still trumped them for latent heat and remained the refrigerant of use in large-scale applications like cold storage and skyscraper air conditioning).

The problems really began when CFC fridges stopped working. As was the norm back then, household appliances were simply crushed and buried – and the gas they contained was allowed to escape.

The development of Freons went hand in hand with household refrigerators hitting the big time. In 1937 there were two million of them in North America, and they remained rare in the rest of the world. By 1955 there were more than 40 million in America, and by 1980 the number worldwide was in its hundreds of millions. These were the numbers in working order. A similar number of older ones had already been disposed of.

In 1974 two chemists working in the United States, the Mexican Mario Molina and his American colleague Sherwood Rowland, hypothesised that CFCs could react with ozone in the high atmosphere. Ozone is a rare form of oxygen, where three atoms form a molecule instead of the normal two. Ozone is actually deadly when breathed in, but up in the high atmosphere, far higher than airliners cruise, a band of ozone acts as a filter for the nasty high-energy ultraviolet radiation that arrives from the Sun. Molina and Rowland had found that this high-energy light was also altering the CFCs that had drifted up there, pinging off chlorines and devastating the protective ozone layer. Eleven years later, work by the British Antarctic Survey confirmed that there was indeed a hole in the ozone layer. This mostly lurked over the south polar region, but there was a similar problem in the north. In a rare act of unanimity – mostly because the solution was an easy one – the world's nations agreed to ban CFCs in what became known as the Montreal Protocol of 1987.

The damaging refrigerants were phased out. Today our fridges probably have PFCs, or perfluorocarbons, in them. They don't damage ozone and at the last count in 2010, almost all the CFCs had gone from the atmosphere. The ozone hole is mending and will be back to how it was in a mere 30 to 40 years. So all good? Well, no. Like CFCs, PFCs and their related fridge fellows are exceptionally potent greenhouse gases. They trap thousands of times more heat energy in the atmosphere than the often-maligned carbon dioxide. As a result, developed nations ensure that all refrigerants are collected when machines are being disposed of, but elsewhere they are left to enter the atmosphere as before.

So could it be that mass refrigeration is doing us more harm than good? That is just what people used to think. They called it frigoriphobia.

NINE

Living in the Chain

I have the body of an eighteen-year old. I keep it in the fridge.

Spike Milligan

The first record of a lobster catching a train was in 1842, when a live New England lobster boarded the express to Chicago. There is no record of how it enjoyed its historic journey, but it was to be its last. It died before reaching Cleveland, whereupon its travelling companions cooked it and put it on ice for the remainder of the journey. Those companions had a contract to fulfil – to provide fresh lobster for a high-rolling Chicago client. The lobster was duly eaten a few hours later, and no doubt enjoyed. But was it actually fresh?

The rail-borne lobster had travelled in the cold chain, a temperature-controlled network that transports perishable foods. Today we live entirely bound up by the global cold chain. As many commentators have pointed out, if malign forces wanted to bring civilisation to its knees, they need not infect the Internet with viruses, launch a military conquest, or interfere with the monetary system. All they need to do is turn off all the fridges. It is often said that no city is more than three meals away from anarchy. If the cold chain broke, society would collapse.

In 1880s Paris, the opposite seemed true. When customers discovered that Omer Decugis, a well-respected and highly successful fruit and vegetable wholesaler, was using a refrigerator to store his produce, the French response was to revolt. Decugis was the first in a long line of fruit merchants – his family company has since grown into an international operation. He had begun his business in the 1850s using the new rail network to transport produce from Marseilles in the south first to Lyon and then Paris. When 'la frigorifobie' hit his Paris business hard, Decugis had to think fast. The outraged customers objected to the apparent lie: Decugis's fruits were sold as 'fresh', but they had been lying around inside his chilled warehouse for days, perhaps weeks – no one knew how long. Decugis's solution had flair. He had the nefarious *frigo* hauled out into the street in front of his shop near Les Halles and smashed it to pieces. With their anger assuaged, the Parisian shoppers returned to buy his produce safe in the knowledge that they were buying fruits and vegetables picked fresh from the fields.

Refrigeration certainly benefited food producers and wholesalers. It allowed them to slow the natural decay process while they shifted their surplus produce in markets far and wide. It also benefited consumers who enjoyed lower prices for food that was in great supply. However, it created a hazy limbo between production and consumption that unnerved many.

In the United States, where the links between food and land are less pronounced than in France and other parts of Europe, the concern was that refrigeration was used by food suppliers to hoard foods to push up the prices and then play the market. Cold storage was also thought to hide the poor quality of foods. Was it even safe to eat refrigerated food? Nobody knew for sure.

These concerns were encapsulated in 337 tons of beef sent to the US Army fighting the Spanish in and around Cuba in 1898. The meat had been packed in Chicago and transported overland in refrigerated railway wagons to New York, then stowed aboard the SS *Manitoba*, a newly purchased transport vessel, for the voyage to the Caribbean. The *Manitoba* was fitted with a refrigerated hold to keep the meat fresh. However, once distributed to the troops on the ground it was soon apparent that something had gone wrong. A Captain Warburton testified at an ensuing federal commission:

> *I ate of the beef which the transport Manitoba brought … It was so bad that it was impossible to swallow it. I had no idea that this meat had been subjected to any chemical process, but believed the beef to have decomposed on account of the lack of proper refrigerating facilities caused by the clogging of the machinery used for the purpose on the Manitoba.*

The 'chemical process' was a reference to the original charge levelled by General Nelson Miles, the supreme commander of the US Army at the time: the Chicago packers had hidden the poor quality of the meat sold to the Army by 'embalming it'. His suggestion was that chemicals had been used to preserve the meat's appearance, when in fact it was rotting away beneath the veneer of freshness.

The investigations failed to find evidence of this macabre deception, and laid the blame in the same place as Captain Warburton. The fridges had not done their job. Today we might think that the supply of meat was handled incompetently. The *Manitoba*'s crew let the refrigerated hold warm up, and there was no ice house or other cold storage for the meat on arrival in Cuba and Puerto Rico – a Tudor ice house would have done the trick. The cold chain had broken. However, in many people's eyes, not least the fuming general's, the so-called Beef Scandal was further evidence of how refrigeration made food and its purveyors untrustworthy.

The general had much cause to be concerned. He was engaged in the Spanish-American War, a short-lived global conflict that saw the United States join the ranks of empire-building states. On fronts in the Caribbean and the Pacific a total of 332 Americans died in action (not counting the skirmishes before war was declared). Almost ten times as many servicemen died from disease. The biggest killers were yellow fever and malaria, but many of those weakened by these diseases were sent to their graves by eating unsafe foods – both refrigerated and canned.

The scandal only enhanced public alarm about refrigerated foods and was an obstacle to food suppliers looking to make use of the technology to sell their produce. The public voice of these concerns was Dr Harvey Wiley, the chief chemist at the US Department of Agriculture. He was instrumental in getting the US Congress to pass laws on food purity in 1906 that were meant to ensure safe practice, and convince the public that the cold chain worked in their favour. However, concerns remained, and to assuage them Wiley began a series of tests designed to check the efficacy of America's cold storage.

By 1910, Wiley had his results. His focus had been on the effect of cold storage on foods. He announced that chilled foods remained perfectly good to eat, ending the debate – for several decades at least – on whether refrigeration kept things 'fresh'.

Let's digest a few food statistics to drive Wiley's point home. It's a hotter than average day (above about 25°C) at some kind of hypothetical medieval marketplace. There is no cold chain and for the sake of argument all the food on sale has been delivered from farms and fishing grounds only minutes away. You cannot get it any fresher than that. The fish and meat will last only a few hours in this heat. Salad vegetables will wilt in less than three days, fruits will be mouldy in less than a week, and carrots and potatoes might last 20 days if stored correctly, but the possibility of rot is never far away. In such conditions, a healthy mixed diet enjoyed in developed societies would be a near impossibility.

Now let's rerun the grocery purchases with the help of a cold chain that is keeping everything at just above freezing point. The meat should be fine for ten days, the fish lasts a week and the salad might last for 20 days. Fruit can stay looking fresh for months, and root vegetables can be kept in cold storage for almost a year.

Refrigeration also boosts the longevity of foods preserved in other ways. For example, if kept just above freezing, a dried jerky will stay good to eat for around three years.

The sharp-eyed and inquisitive will have begun to question these figures, which appear longer than we are used to storing foods at home. In addition, any lawyers reading this would advise us not to use the figures as a guide to food safety. There are two reasons for this. First, a domestic fridge runs a little warmer – around 4°C – so its preservation abilities are slightly diminished. Second, these times relate to the entire period the foods are in the cold chain. A good deal of that ticks away as the food is transported to your supermarket, reducing the window of freshness by the time you can get it home.

The modern cook seldom interrogates the word 'fresh' today. Whatever goes into the refrigerator looking (and tasting) as though it has just left the field or slaughterhouse should come out again looking just the same. Normal refrigeration at around 3°C or 4°C slows the decay process driven by all-pervading germs and moulds (although it won't quite stop it completely).

Unfettered by any form of preservation, these bugs start to alter the food. Basically, they start to digest it for themselves. That changes the food's colour, makes it soft and even slimy, and it begins to whiff in a way that we are hardwired to regard as nasty, not least because it indicates that it will make us sick.

Traditional preservation techniques – drying, salting and pickling – kill off these bugs. The food will now be safe to eat for much longer, refrigeration or no, but the process will have altered the flavours and consistency of the foods. This is no bad thing in many cases. Perhaps some would agree that smoking a salmon or pickling a cucumber makes it even tastier. However, cold preservation does it differently. The fridge is a kitchen time machine that puts the food inside into slow-mo. The metabolism of the germs on the food (currently in harmless quantities) grinds almost to a halt, lengthening the time that the food is 'fresh'.

Deep freezing makes decay stop entirely – and the germs slowly starve to death in an ice-crystal tomb. Below freezing, storage times grow from days and weeks to months and years. However, the process of making food this cold – and frozen solid – can also damage the structural integrity of most foods. So once defrosted it is no longer seen as 'fresh' either – although it is free of decay. The man who did the most to overcome the 'fresh-frozen' problem had a famous name – Birdseye. More on him later.

Even above freezing, different foods need to be handled in different ways for them to arrive 'fresh' at market. For example, fruit is damaged by excessive cold and needs to be stored at a perfect humidity to keep it tip top. All that know-how took many years to perfect, and the pioneers of the cold chain battled politics, science and economics to make it work. The risks were great – not least the considerable hazards of rotten foods – but the rewards were greater.

*

In July 1875, shipping magnate Thomas Sutcliffe Mort got up to speak at a congress of the Agricultural Society of New South Wales. His topic was the supply and demand of Australian meat. At that point there were 1.7 million people living in the main Australian colonies – he did not count Western Australia and the Northern Territory, both relatively empty at the time. The same four regions had 5.6 million cattle and 47.8 million sheep. He allocated every Australian, man, woman and child, 430g (15oz) of meat a day (that's about four quarter-pounders, and although some might regard that as a reasonable amount, it's more than seems sensible in modern terms). Mort then calculated that even this maximal meat diet would leave Australia with 275,000 tonnes of spare meat every year. What to do? Mort had the answer: ship it to Britain where the growing masses at the heart of empire seldom got a hint of quality fresh meat.

The Times (of London) had recently reported that very problem and its apparent solution: 'There are parts of the world adapted for maintaining vast herds of cattle and sheep, and adapted for no other purpose, and from these the more densely populated countries might be permanently supplied with sustenance.' Mort was going to give it a try.

All he needed to do was transport thousands of tons of cattle and sheep the best part of 17,000km (10,500 miles) and keep them fresh along the way. One obvious method was to transport live animals. However, just getting the animals to the dockside would be hard enough. They were transported in rail cars often without food or water, and arrived in a very poor state. In the United States, railmen were also replacing cowboys as the people charged with moving cattle from the Great Plains to the meat-packing hub at Chicago. Even when supplied with ample food and water and rested as often as possible, railway-transported cattle were just too frightened by the ordeal to eat. They emerged at the other end about 40kg lighter than when they boarded.

The answer was to slaughter animals at the start of the supply chain and transport 'dead meat' kept fresh by refrigeration.

Transatlantic liners had kept food for their passengers fresh on ice since the 1850s, but none of it was expected to last longer than the crossing. Dressed American beef was already being imported to Europe on board ice ships, and the fridge pioneer James Harrison had attempted to show that Australian meat could travel in the same way in 1873. It was a total disaster. The ice melted long before the precious cargo made landfall. Harrison left the refrigeration game after that and went back to the safety of journalism, preferring to write about the deeds of others than risk any further deeds himself.*

*Who can blame him? Not me.

Thomas Mort's plan was to use mechanical refrigeration instead of ice. In the 1860s he had diversified his shipping business to include among other things the New South Wales Fresh Food and Ice Company. He barely made a profit from it but the business worked. One of his party tricks was to freeze fly-blown cheese riddled with maggots. He then had it defrosted in the sunshine and gathered an audience to watch the maggots reanimate from their icy torpor.

In 1877 Mort was ready to combine his ice and shipping business in the form of the *Northam*, a barge converted into a floating refrigerator. In July, midwinter in Sydney, the *Northam* was made ready to carry an initial shipment of lamb to London, and Mort had future plans to shift fish, milk and even live oysters to the other side of the world.

Everything appeared set. The *Northam*'s hold was a cool 11°C when the meat went in, and the refrigeration machinery cooled that down to about 1.5°. In 1874, Louis Pasteur, the famed French chemist who had formulated the germ theory of disease and decay the decade before, had announced that meat kept at this kind of temperature would stay good to eat for 50 days. Nevertheless it was not clear if the *Northam*'s temperature was going to be sufficiently chilly for a 90-day voyage. Mort was just going to have to risk it.

But he never got the chance. The ammonia-compression technology let him down before the ship had even left the quayside. With the hold heating up, there was nothing else for it but to unload the meat. Mort's dream was over and he was left with a sizeable financial worry. He died the following year.

Mort was not the first to suggest refrigeration ships. In 1868, the government of Argentina had offered a cash prize of 40,000 French francs (about £120,000 in today's money) to anyone who could develop a means of refrigerating the beef being produced in ever larger quantities on the pampas. The prize was never awarded, but the concept caught the eye of Charles Tellier, a French engineer who had developed a refrigeration system for cooling rooms as opposed to making ice.

In 1876, Tellier raised funds by public subscription to do the equivalent of taking 'coals to Newcastle', or more pertinently 'ice to the Eskimos': Tellier was going to transport fresh beef to Argentina. To be fair, the voyage of his newly refurbished and refrigerated vessel, the *Frigorifique*, was to be a proof of concept. If the meat survived the crossing, then the importation of fresh South American beef could begin in the other direction.

The *Frigorifique* set off from Rouen at the start of September with a cold store of steaks. A storm in the Bay of Biscay threw it off course, and the ship sought refuge in Lisbon. The King of Portugal even took an interest in the cargo, which is reported to have survived the three-day deluge untainted. Tellier's backers declared the experiment a success, and began to dump their shares at a high price – but Tellier insisted that the ship cross the Atlantic. The *Frigorifique* arrived in Montevideo on 23 December, and nipped over to Buenos Aires on Christmas Day. By now the steaks were 105 days old. When unloaded they were declared to be 'excellent, bloody and savoury' by the welcoming dignitaries. 'Hurray, a thousand times for the revolutions of science and capital,' enthused *La Liberté*, a Buenos Aires newspaper. 'The dawn of a new day rises for La Plata.'

True enough, frigorificos, or refrigerated meat-packing plants, began to spring up in the River Plate's ports. 'Dead meat' from the Americas was on its way to Europe.

*

Vessels equipped for refrigeration became known as 'reefer ships'. The technology was perfected by Scottish brothers Henry and John Bell and their English partner James Coleman in the 1880s. Their system (they also got a little help with the design from Lord Kelvin himself) worked differently from refrigeration on land. Steam-powered engines on board the reefer were used to compress air, which was then blown into the holds, packed with perishables. The compressed air expanded rapidly and cooled, plunging the temperature of the cargo to below freezing.

An early success of a Bell-Coleman reefer was the *Dunedin*, which at first glance looked like an outmoded clipper, powered by sail. However, in 1882 this vessel made a voyage that entered New Zealand folk memory. It was loaded with 4,331 sheep carcasses, 598 lambs, 22 pigs, 246 barrels of butter, 2,226 sheep tongues and a mixed bag of hares, pheasants and poultry. The meat was frozen in port and stayed frozen until it arrived in London 98 days later. Refrigeration provided a crucial link between the farms of distant New Zealand and the rest of the world that exists to this day.

A typical reefer ship consumed 2.5 tonnes of coal a day but was able to keep the cargo frozen all the way through the tropical zones. The shipyards of Glasgow built most of the ships, and by 1902 there were 460 reefers at sea, carrying in the region of a million tonnes of refrigerated produce at any one time from the Americas, Australia and New Zealand.

The arrival of industrial refrigeration in the 1880s put the food industry into the hands of the owners of the cold chain. Over the decades, the frigorificos of North and South America were bought up by a handful of Chicago businessmen, who formed a cartel known as the Beef Trust. Admittedly, the Trust's product was of excellent quality compared with locally produced meats. Despite having travelled thousands of miles and being considerably older, it still looked and tasted fresher. However, by the early 1900s, the Chicago cabal was able to set the price paid to ranchers across both continents, and ensure a healthy profit from sales in Europe. Only in the 1920s, when enough reefers sailing from beyond their reach – chiefly Australia and New Zealand – could match the quality and quantity of supply did the Trust's international influence begin to wane.

A similar process overtook the trade in tropical fruits. Set up in Boston in 1899, the United Fruit Company developed a near monopoly on bananas and other fruits from Central and South America that would last into the 1960s. The company became known as 'El Pulpo' (the Octopus) because its tentacles appeared to reach into every aspect of economic and political life in Latin America. It is a commonly held belief that United Fruit was the hidden hand behind the repeated military coups that befell the region's 'banana republics' throughout this period.

The fruits were carried in reefers painted white to reflect the warmth of the tropical sun. Many banana boats offered passenger services too, and the ships became known as the Great White Fleet. People in no particular hurry would cruise lazily around the Caribbean as the vessel collected fruit cargoes. The cruise industry is rather different now, although not completely. The passengers are still happy to go nowhere fast, and the ships are still white, but there is little room in today's floating palaces for fruit cargo.

Keeping fruit fresh requires a different kind of cold storage than that used for transporting meat. If fruits are not chilled while being moved they will ripen en route and be unsellable by the time they get to market. However, soft fruits are also intolerant of the kind of cold used for other perishables. For example, if bananas are stored below about 8°C for any length of time, their skins turn black. The fruit inside is fine, but customers won't buy it because they assume it is going to be old and bruised.

Fruits are also preserved best in a controlled humidity. If it's too arid, they'll dry out; if it's too dank, they'll rot. Therefore, the holds of banana boats were fitted with fans that blew cool air over the fruits. The fans could be switched to warm the holds when ships entered wintry conditions.

Trains, or at least their carriages, could be reefers too. For much of their history they were cooled by ice, and in Europe, where distances were shorter and the cold chain much more ephemeral, fruits were simply transported in open-sided vehicles to create a cleansing and cooling breeze. Australia began dabbling with reefer cars in the 1890s, but it was Central California, America's fruit basket, that showed the rest of the world how it was done.

Like United Fruits and the Beef Trust before them, the railroad companies saw the opportunity afforded by refrigeration. The Union Pacific and Southern Pacific went into partnership and set up the Pacific Fruit Express (PFE) in 1906. The PFE's 6,000 reefers carried fruits across the country, chilled with ice that had to be restocked nine times during a cross-country trip.

As well as being the biggest fruit company in North America, the PFE became one of the world's biggest ice producers, as it built ice plants across the rail network to service the fruit's insatiable need for cool. This ice supply chain was in action well into the 1970s, when mechanically cooled rail cars were finally adopted in North America.

The mechanical reefer had taken a long time to take over. Its inception occurred during a game of golf in Minneapolis on a hot summer's day in 1938. One of the players was bemoaning the fact that his shipment of chickens had spoiled after a truck had broken down en route from Chicago. Another player was Joseph Numero, a businessman who provided sound equipment for local movie theatres. The golfers challenged Numero to find a way of incorporating refrigeration into a truck. Numero was an entrepreneur and no engineer, but he already knew someone who could do it. 'We can build a unit for you in 30 days,' said Joe.

It took a little longer than that, but by 1939 Joe and his engineering partner Frederick McKinley Jones had patented just such a machine, known as the Model A. The pair set up a company, Thermo King, to manufacture and market the units. Thermo King is now a global player in the cold chain. More Thermo King models soon followed, but the well-established infrastructure of ice plants prevented them from making much of a dent on the cold-chain status quo. Thermo King was not the first to try mechanically refrigerating trucks, but Fred Jones's designs proved more resilient than earlier attempts. One of his innovations was to make the unit more lightweight so it could be relocated from under the truck to the top, where it was less likely to become clogged with dirt.

Fred and Joe's first big success came in 1942, when Thermo King provided the American military with reefer trucks. These were used in the wartime supply chains that took fresh food up to the battle fronts, where no iceman would dare to tread.

By the late 1940s Thermo King was making inroads into railroad stock, producing refrigerated box-cars to replace the block-iced reefers. However, progress remained slow. In the 1950s, Thermo King diversified into shipping and repurposed its technology to provide air conditioning on buses, but the real turning point came in 1959, with the introduction of diesel-powered units. These lasted five times longer than petrol-driven units. Reefer trucks were now able to compete with rail transport, and refrigeration units fixed to the front of semi-trailers began to become the familiar sight they are today.

In 1991, Fred Jones and Joe Numero were posthumously awarded the National Medal of Technology by President George H.W. Bush. Jones was the first African American to receive this award. From Model A onwards, the links that Jones added to the cold chain connected it to the modern world. Thanks to him, Joe and those golfers sweating on the fairway, we have become accustomed to ripe fruits and raw meats travelling great distances to reach our tables. The produce is perhaps out of season – although few of us really know what this is – but it is always fresh.

Something else that had changed was the idea of freshness itself. In a meeting to discuss the veracity of the term 'fresh' on food packaging in 2000, a US food-industry insider is recorded to have said: 'Fresh is not a measurement. Fresh is a state of being.'

An article in the journal *Ice* in 1909 had a more sniffy attitude: 'Unfortunately, the full richness of flavour is found only in the fruit that matures upon the vine, or tree, and thus New York, while eating ripe California fruit, never has known how good California fruit ought rightfully to be.' What was true then is true now, the world over. Domestic refrigeration has changed food forever.

❋

In 1956 Kathleen Ann Smallzreid wrote a book, *The Everlasting Pleasure: Influences on America's Kitchens, Cooks, and Cookery, from 1565 to the Year 2000.* That's quite an ambitious title, but she makes it work with this excerpt:

> *When a housewife returns from the supermarket and whisks things into her refrigerator and closes the door, she has closed the door of the springhouse, the milk and butter pantry, the root cellar, the cheese room, the smokehouse and the covered well. At the same time she has turned her back on the preserving kettle, the pickling crock, the pudding bag, the vinegar barrel. As for the icehouse, the ice wagon, she has put them behind her too. Ice does not make the storage box cold. Instead the box makes ice for all her needs.*

While the cold chain began as a means to move surplus produce further, faster and for longer, it has ended up changing our relationship with food entirely. Domestic refrigerators gave families their own cold-chain terminal – first in America, and later the world. It was the place where Californian melons that had rattled over the Rockies and across the Great Plains in iced railway wagons could meet up with steaks from the Chicago packing houses, cod from the Grand Banks and bananas from Cartagena.

The cold chain had broadened the range of foods available at the grocery store and now people were offered somewhere to keep it all. The home refrigerator was sold with a triple whammy: it stopped all that exotic food from spoiling, it saved you time and it made a personal supply of ice. That last one was the initial draw for customers, because it was something that an icebox couldn't do, but it was the time saving that really made the difference. Housewives (for it was their refrigerator) were freed from buying foods every day and they were no longer compelled to slave away preserving perishables. That created a new opportunity for the food industry – why not sell families a kitchenful of food all at once? In other words, create a supermarket.

Although it would sell a great deal of non-perishables as well, the supermarket was impossible before refrigeration. This was not just because customers needed refrigerators to store all the food they had bought, but because the store itself needed to keep its perishables in giant ones (the supermarket is also a product of the car age; another cost-cutting idea was that customers transported all their purchases themselves).

The world's first attempt at a supermarket was the Astor Market set up by Vincent Astor, a scion of the philanthropic dynasty. This opened in the Upper West Side of Manhattan in 1915. The business model was as you'd expect: a large selection of products available at low prices due to the economies of its scale. But the Astor Market was too early. Not enough people came, and the business folded in 1917. Refrigerators were more expensive than cars back then, and few people had either.

Customers at the Astor Market were expected to queue up at different stalls around the building to get their different items. The idea of a store where customers helped themselves belonged to Clarence Saunders, an American grocer. Saunders was the founder of the Piggly Wiggly grocery chain (the name has stuck to this day). The first store opened in Memphis, Tennessee, in 1916, and saw customers entering via turnstiles as if they were visiting a funfair or boarding the subway. They got out again by paying for everything once at a checkout. The modern shopper knows exactly how it works.

Piggly Wiggly only sold non-perishables to begin with. A true supermarket sells it all – fresh produce along with the rest – and there are several claims on the title of 'World's First Supermarket'. One that certainly met the bill was the King Kullen chain that began in Long Island in the 1930s. King Kullen 'piled it high and sold it low', and provided ample parking for its customers.

King Kullen and the other pioneering chains started up during the Great Depression and offered the economies that Americans were looking for. In the last 20 years, supermarkets have invaded even the most stalwart gastronome nations. The reason for the slower uptake in Europe was twofold. Firstly, the convenience factor was less compelling. Residents of continental cities were seldom more than a few minutes from a grocery store. Secondly, there was a stronger link with food. People knew what ripe fruit and vegetables and fresh meat tasted like, and cold-stored products did not cut it. Nevertheless, the variety offered by large edge-of-town hypermarkets has proven hard to beat.

A lot has changed inside supermarkets since the 1930s, but really nothing has. Supermarkets have swelled to incredible sizes and sell a great deal more than just food. But in essence they are the same as they have always been: teeming nodes in the cold chain where refrigerated produce is dispersed on the last leg of its journey.

The supermarket has compressed our grocery purchases into the 'big shop', which takes place once a week or fortnight. In the UK, which in recent years has forged a cutting edge in the business of supermarkets, millions do their big shop online and it is delivered from so-called 'dark stores'. A dark store is a supermarket like any other, except there are no checkouts, no marketing displays and not much in the way of lighting. There are also no customers, just staff working around the clock to assemble online orders for delivery in refrigerated vans.

Allowing someone else to choose your groceries, especially your favourite soft fruits and cuts of meat, requires a degree of trust. The online shopper is happy to conclude that one punnet of peaches is the same as another, and one packet of lamb chops looks as tasty as the next. Above all, online shoppers take it as read that food will be fresh on arrival and stay fresh for a known time.

For many of us, getting hands-on with food shopping, sniffing a melon, squeezing an apple or nibbling a sample of cheese is now a leisure activity, something to be enjoyed on a care-free weekend trip to an ever-so-rustic market. The rest of the time we just load up the basket (virtual or otherwise) and move on. There is no time or money to spare.

As that *Ice* journal writer lamented in 1909, the cold chain creates a division between food production and consumption. Our first encounters with food are in the supermarket. That's just where it comes from. From the outset, cold-chain operators worked hard for their products to arrive looking just right at the point of sale. That is no bad thing. The quality, by some measures at least, had to go up. If it didn't then a superior rival supply would soon replace it.

In the case of mass-produced fruits, it was a final chapter in a long story that began in the Middle Ages. It is no accident that the wicked and vain queen chose an apple to kill Snow White, as the story goes. Until the advent of the cold chain, the apples and other fruits that made it to market were often unripe, having been picked too early. Their bitter taste reminded people of poison, a very real threat in medieval times. To compound the issue, apples left to ripen erupted with maggots – more probably moth caterpillars, but the afflicted would not thank you for the clarification. To the pre-Enlightenment mind, apples were certainly the Devil's fruit.

Fresh fruits eventually found a place on Renaissance dining tables for decoration. Then as now, people frequently judged the freshness and purity of fruits by what they looked like rather than how they tasted.

There are of course valid concerns about food miles and the energy resources used to stock our shelves with foods harvested from across the world. In some cases the statistics are stark. Two-thirds of US land and 70 per cent of its grain is diverted from supplying the human diet to rearing livestock. Adding that all up means that to make 1kg of beef, farmers must use 100 times the amount of water it takes to make 1kg of vegetables. It is pretty clear that the demand for such intensively produced food cannot be met with the natural supply of land and water. However, the subject is fraught with complexity. Locally produced vine tomatoes grown under lights a few miles from a shop will probably require at least as much energy to produce (especially in winter) as the green beans rushed in by air from the tropics.

Whether these growing concerns can be met by the cold chain and food industry at large remains to be seen. Its primary concern is selling food, and it has many a willing buyer, as was already obvious in 1910, when the *New York Tribune* wrote:

> *The effects of the* [refrigeration] *industry have become thoroughly entrenched in civilized mans' habits of living. He expects to eat eggs all the year round, while his grandmother used them in winter only for cake baking. Likewise apples in May and June are taken as a matter of course, whereas preceding generations did without.*

Today's civilised man (and woman) expects much the same, and to get it their eggs are coated with a sheen of mineral oil to seal the porous shells so they do not go rotten in storage. Similarly, their apple's natural waxy peel is coated with a food-grade paraffin wax to provide a barrier to moulds as it begins its long journey through the chain.

These and other tricks of the trade obfuscate the freshness of perishable produce, but there is no hiding its age. A strict regime that records packing times and displays consumption dates sees to that. The 'use by', 'sell by' and 'best before' dates all have nuanced meanings, but they are the timetable that our food conforms to. The timeline errs very much on the side of caution. Think back to those French steaks barrelling through the Atlantic for three months on board Tellier's *Frigorifique*. Their use by dates would have passed long before they were served up on the quays of the River Plate. Snooty foodies might wrinkle their noses at a long-gone 'best before' – not because the food smells foul, but because it does not smell or taste off at all. Admittedly there is some latitude built into the system.

However, one of the chief drivers behind these dates was the discovery of listeria. This bacterium was first identified in the 1920s and was named after Joseph Lister. Lister was a nineteenth-century Scottish pioneer in the use of antiseptics in medicine, and it is perhaps unfortunate that a man who saved so many lives is honoured in the form of a deadly germ. Listeria is a tough old bug, and unlike most other germs will keep growing slowly but surely, even when refrigerated. It's a one or two in a million kind of disease, but infections that lead to a case of listeriosis, a disease of the nervous system, have a 25 per cent mortality rate.

In the United States, incidences appear to be on the decline as better hygiene standards have taken root along the cold chain. In Europe, where domestic refrigeration is a newer habit, the reverse is true, but the risks remain small.

Those risks arise not from raw ingredients destined to be cooked before eating, but from ready-to-eat fare, like cooked meats, unpasteurised cheeses and ground-grown fruits. The most at risk are pregnant women and people with immunity disorders. It is easy to be alarmist about listeria, but the use-by dates are just the last line in an impressive defence against this and other bacteria. Don't be alarmed. Always read the label.

*

Frozen food does not suffer the same issues with germs. Water freezes at 0°C but food won't go solid until it is about −2°C all the way through. A few bugs will still multiply at a sub-zero temperature as long as there is some liquid water around, but once the food becomes solid, their growth stops. A deep freezer will take food to below −9.5°C, which is the point where even salted liquids will freeze and the spoilage process becomes negligible. Don't be fooled though. The germs are still there and will come back to life once the food is thawed just like Thomas Mort's performing maggots.

Using the home refrigerator to freeze foods became popular in the 1950s (again chiefly in America). Supermarkets gave consumers the opportunity to buy their favourite foods in bulk at low prices. But they needed somewhere to store it all. If they could get everything into their regular refrigerator, they would struggle to eat it before it became unsafe. The battle to convince people that food stored in a fridge stayed fresh and good was by this time well and truly won. It was therefore a relatively simple next step to convince people that pre-frozen foods had not lost much quality either. The convenience and economy of freezing food was a no-brainer to a society already wedded to refrigeration.

Companies began offering stand-alone freezers. They worked using precisely the same method as the kitchen refrigerator, but were generally more rough-and-ready, hefty chests built for the cellar, garage or similar. While the kitchen fridge was finessed with fashionable looks and internal dispensers, trays and other gizmos, the freezer was just a storage box. There was not even a light that flicked on when the door opened.

Why that should be the case has become something of a riddle. Why is there a light in the fridge but not the freezer?

Fine minds have come up with all kinds of seemingly sensible answers. A typical guess is that the bulb would become a heating element, melting foods and promoting bacterial growth. Sure, a bulb is warm, but it would only be heating foods when the door was open. Keeping the door of the freezer open is a more effective method of melting food and promoting bacterial growth than using a light bulb. Another guess seems a bit more intelligent. A glass bulb would be at sub-zero temperatures most of the time. Turning it on would result in a rapid rise in temperature as the filament brightened to white hot. Surely that thermal shock would crack the glass? Sounds plausible, but toughened glass would solve that problem. The real answer is a lot more prosaic. A light bulb and its attendant electrical supply would just cost too much. Then as now, we didn't open the freezer very often, so any bulb would be idle for most of its lifetime. When we did open the door we knew what we wanted and had a pretty good idea of where it was. No one stared into the freezer wondering what to eat. Finally, we could just turn on the room light if needed. Nevertheless, a modern freezer compartment will probably have a light now. The fixed costs of it are small enough to make it worth the added wow factor in a competitive market.

Such technological trivia was not the major obstacle to the development of deep freezing. The biggest problem was that foods that had been frozen generally thawed out into various grades of mush. Meat survived best, and freezing was even promoted as a means of tenderising it, but a frozen steak could not really match traditionally dried and dressed cuts. Dried foods were largely unaffected by freezing, but there was seldom much call to freeze this stuff anyway. Fresh and cooked fish, fruits and vegetables were invariably ruined.

The reason was the rate of freezing. A slow drop in temperature allows individual ice crystals to grow to a large size. As they enlarge, the crystals rupture all the delicate cellular structures that make up food. Once thawed again the food is flaccid, seeping juices, and has become tasteless in the process. The solution was a method of rapid freezing, and it was Clarence Birdseye who figured it out.

Clarence was an American but is most famous today in the UK under the name Captain Birdseye, a salty sea dog intent on delivering frozen foods to the nation. Other Europeans know him as Captain Igloo. In truth Clarence knew a thing or two about igloos, but was not a seaman. He did have many jobs – taxidermist, fur trapper and entomologist – but was a definite landlubber.

Born ashore in Brooklyn, New York City, in 1886, Birdseye was a college dropout who moved out west to work for the US Department of Agriculture in New Mexico. His first job in 1908 mainly involved killing coyotes, but by 1910 he had become an assistant to Willard van Orsdel King, a government scientist who did much to figure out the first effective controls of malaria-carrying mosquitoes. Birdseye travelled around the West trapping and preserving small mammals as part of a research project into their insect parasites. It was through this work in 1911 that Birdseye helped figure out that Rocky Mountain Spotted Fever was spread by bites from a tick. At the time this disease, a form of typhus, was poorly understood and often fatal (several later researchers died from it). Fortunately for Birdseye – and for the future of the food industry – he moved on to a new project in 1912. He went to Labrador, then part of the grandly named Dominion of Newfoundland, to get rich as a fur trapper. However, once there he got rather sidetracked.

The local Inuits showed him how to ice fish, and Birdseye saw that the catch froze almost instantly in the sub-Arctic conditions. Once thawed the fish looked and tasted perfect. The freezing occurred so fast at –40°C, or thereabouts, that only tiny ice crystals formed, crystals that were smaller than an individual cell and therefore did no damage to them.

By 1922 Birdseye had the backing to begin developing this kind of flash-freezing fish on an industrial scale. It took until 1925, and the inevitable brush with bankruptcy, to get something that worked – which eventually became the Birdseye Multiplate Freezer, no less. The fresh fish was packed flat to maximise its surface area. It was then placed in cartons that were fed onto a steel conveyer. The metal surface was chilled to −43°C using a refrigerated brine that was circulated underneath it. A similar system was cooling the roof above the conveyor, which was pushed down onto the cartons as they passed through. The machinery was frozen as rapidly as any Inuit fishing party could hope for. There have been advances since then, but the automated flash-freezing systems used today work in much the same way.

Birdseye set up the General Seafood Corporation in Gloucester, Massachusetts, and the rest is corporate history. In Europe, the Birdseye brand, which also freezes meats and vegetables, found its way to Unilever (and has recently moved into other hands). Unilever are also the largest ice-cream producer in the world, an empire that began with the acquisition of the UK company Wall's in 1922. Strangely, Wall's was originally a butchery business, a venerable 'purveyor of pork', whose customers included King George IV. However, its owner Thomas Wall had spent the previous ten years exploring ways of maintaining staff and turnover in summer when trade in his famous sausages and pies dropped away. His answer was to stock his ample cold storage with ice cream (made smooth and soft with a large dollop of pig fat from the butchery), something that Unilever, then known as Lever Brothers, put into action. The company never looked back (although pig fat is not an ingredient today). Today Unilever sells £4 billion of ice cream a year and owns everything from America's Ben & Jerry's and Ecuador's Pingüino to Austria's Eskimo ice cream and Australia's Streets brand.

In North America, Birdseye is now owned by Pinnacle Foods, but spent many years as part of the Kraft Foods Group. In 2008, Kraft, a major figure in the cold chain, did something rather unusual. It moved into a subterranean lair. This might sound like something that Mr Burns from *The Simpsons* might do, not least because the facility is in Springfield, Missouri, but the vast cave is an excellent place to keep things cold. Springfield Underground is a limestone mine buried 30 metres (100 ft) under the ground and has more than two square kilometres of storage space. The natural temperature is an unwavering 14°C down there, morning, noon and night, but 50,000 litres of brine coolant pumped through its vast caverns bring it down to a steady 1°C. If a James Bond villain ever needed a refrigerator, they'd now know where to go.

The processed cheese, sandwich spreads and jelly desserts that lurk in the cave come and go in refrigerated containers hauled by trucks, big rigs, artics and semis. The taste for American cheeses does not spread far overseas, but in theory the containers coming out of that cave could travel the world. The modern cold chain is now containerised just like everything else, and refrigerated cargoes can move from truck to train to ship and back again without the need to touch what's inside. Individual refrigerated containers are able to maintain their own internal climates, preset before departure to suit whatever is inside. They were developed in the late 1970s, and largely thanks to the Queen of Cool, otherwise known as Barbara Pratt.

As an employee of Maersk, one of the world's shipping behemoths, it was Queen Barbara's job to live inside a refrigerated container for weeks on end. The container was a bit special. Maersk called it the Sea-Land Mobile Research Lab. On the outside it was like any container apart from the bulletproof windows (it was frequently left in ports, which can be unsavoury, lawless places). On the inside there was a small living and sleeping area at one end. At the other end was a laboratory. The idea was to create the same conditions as inside a container at sea, on the road, or on railway tracks. Therefore once Barbara and her fellow scientists were inside, the doors could not be opened again until an experiment was completed. The lab was easy enough to transport to different test sites – after all it was in a shipping container. Barbara and her team travelled around with the lab, but most of their work was carried out in one location – a port terminal, farm or factory – chosen for its particular conditions. For the best part of seven years, Barbara used the lab to monitor humidity, temperature and air flow inside a container and how they affected different foods that would be carried inside. Her findings led to design tweaks and guidelines on how to pack food for containerisation that exist to this day.

The Sea-Land lab was one of many places where science and refrigeration met. In this case the refrigeration was a tool for preserving foods and other natural stuff. However, by ramping up the cooling, refrigeration can be used as a tool to probe the very stuff of nature.

TEN

Deep Cold

Sir James Dewar
Is a better man than you are.
None of you asses
Can liquify gases.

Anonymous clerihew, 20th century

George Mallory, the British mountaineer, is said to have explained his desire to climb Mount Everest in 1924 with three famous words: 'Because it's there.' In the event he stayed there for the next 75 years, his body lying frozen in the Death Zone some 600 metres (2,000 ft) below Everest's icy summit until it was located in 1999. His companion Andrew 'Sandy' Irvine has never been found. No one knows if they made it all the way to the top.

The temperatures at the top of Everest frequently drop to a terrifying -40°C, an unimaginable cold that is at the limits of human endurance. It was even colder on the 'winter journey' of 1911 when three British explorers set out into the Antarctic night. Henry 'Birdie' Bowers, Edward Adrian Wilson and Apsley Cherry-Garrard spent five weeks walking through the darkness at temperatures of -57°C. Their aim was to collect the first specimens of emperor penguin eggs from the breeding grounds of Cape Crozier. Amazingly they succeeded, with all three making it back to base, despite losing their tent for 36 hours and being forced to sleep in a snowdrift. Wilson and Bowers died eight months later with Robert Scott as they returned from the South Pole itself.

The temperatures endured by these men were beyond human experience. Body tissue does not have the ability to feel such cold. It just slowly dies. But scientists knew that colder temperatures were possible, and the same spirit of adventure gripped their exploration of a cold so deep that it could only be imagined, never touched. In the 1890s, James Dewar set his sights on 'Mount Hydrogen'. Mixing his metaphors a bit too much, he was aiming to go lower than had ever been thought possible by condensing hydrogen gas, the lightest and most ephemeral of substances. What he would find down there, nobody knew. And it was going to take courage; it could kill him.

❋

In the 1780s, the pneumatic chemists, those inquisitors of gas, had gone a bit wet. The Dutchman Martinus van Marum had confirmed the suspicions of many by showing that putting a gas under a higher and higher pressure did not result in it just filling a smaller and smaller space ad infinitum. Eventually it became a liquid. He achieved this feat using ammonia, helpfully isolated by Joseph Priestley a few years before, and of course the same fluid that would one day surge through the world's fridges.

However, the temperature changes that occurred were simple data points at this stage. The focus was on the physical changes. In the 1820s, Michael Faraday succeeded in liquefying carbon dioxide and chlorine by chemical means and by chance rather than design. His apparatus was capable of evolving these gases at such high pressure that they condensed in small amounts in the collection vessels.

By 1834, a Frenchman had built an enhanced compressor so powerful that it could produce liquid carbon dioxide in much more useful quantities. We say Frenchman because his identity is to this day often attributed to the wrong man. Until a bit of detective work among the patent libraries in Paris proved otherwise, the person in question was identified as Charles Thilorier, a student of chemistry at the city's polytechnic around this time. However, in 2003, the actual innovator was confirmed as Adrien Thilorier, who it appears had the bad habit of signing his work simply with his surname.

So, now restored to his rightful place in history, Adrien Thilorier had built a compressor capable of producing pressures of 1,000 atmospheres. The device fused the highly geared clockwork used by watchmakers with the brute strength of steam-engine ironwork. Its power was not to be taken lightly. During one demonstration of the equipment, an iron gas tank exploded, mangling the operator's legs. He had one amputated shortly after the accident and was killed by the resulting infection within days.

Nevertheless, the Thilorier machine could do more than liquefy carbon dioxide. It could freeze it. Solid carbon dioxide was first seen inside Thilorier's scary pump in 1835. First described simply as a fluffy snow, it has since become known as dry ice. In normal atmospheric conditions, dry ice does not melt. It sublimates, which means that it goes straight from a solid form to a gaseous state, bypassing the liquid phase altogether. Chips of dry ice plopped into regular water will bubble into a rather theatrical cloud of mist, which flows up and over its vessel and spreads across the floor. There was a time when no magic show was complete without it. When Merlin or some other wizard pops up on screen he is also using it to give his potions that extra spookiness. We must assume that the subterranean lair of Camelot's wizard was rather cold, because dry ice forms at -78.5°C.

Back in the real world, Michael Faraday seized upon the work of Thilorier to push the boundaries of cold further. In 1838 he made a public performance at the Royal Institution in which he recorded a temperature of -110°C. To achieve that great cold he used a 'Thilorier mixture', a frigorific concoction of ether and dry ice. Faraday's plan was to go on and use this mixture to see if he could liquefy nitrogen, oxygen and the other gases that had so dominated the world of chemistry for the last 80 years. However, Faraday then suffered a mental collapse. It appears that a combination of overwork and a probable case of mercury poisoning sent him over the edge.

Faraday took until 1844 to recover, and he then returned to his work on compressing and cooling gases into liquids. By the following year he had succeeded in most cases, but was unable to make headway with oxygen, hydrogen and nitrogen. He dubbed them the 'permanent gases' along with a few others with which he had no joy. Faraday gave up at this point and went back to revolutionising the field of electromagnetism.

Faraday had reached the limits of his equipment. A few others had had a go at extending those limits. One rather ingenious idea was had by Georges Aimé, who used the pressure of the deep ocean to boost his hand-cranked pump. He reports compressing iron cylinders of nitrogen and oxygen as much as his equipment would allow, then lowering them into the sea, using the great weight of the water to squeeze the gases further. Aimé says that he lowered them deep enough to achieve an unprecedented pressure of 220 atmospheres – probably in the order of 1,500 metres (5,000 ft) down. However, there was no evidence of liquefaction. In 1846, the Austrian Johannes Natterer threw engineering at the problem and built a tougher version of Thilorier's pump. He reported achieving pressures of 600 atmospheres. But still no liquefaction.

Within a few short years, a discovery arrived to break the apparent deadlock. We can now welcome back James Joule and William Thomson. In 1852 they presented their findings on the cooling effects of expanding gases, the so-called Joule-Thomson Effect.

Remember, this effect is what is cooling your fridge. When you allow a gas to expand rapidly by pumping it through a nozzle, it not only reduces in pressure – it also cools down. The effect showed that the clean lines of the gas laws, those linear relationships between pressure, volume and temperature began to bend when tested at the extreme end of natural conditions. Several researchers followed the wavering relationships, plotting their twists and turns as gases cooled and condensed into liquids. Many great names had a hand in this work, such as James Maxwell Clerk and Ludwig Boltzmann, two mighty intellects that helped to render nature's forces and matter into mathematical equations. However, a less well-known name, Johannes Diderik van der Waals, gave the best explanation as to what was going on. In 1873, this Dutch academic presented his idea that pure gas and pure liquid are really the same thing, only at extreme ends of a continuum in which the two states of matter are more often coexisting.

In a solid, the constituent units of a substance – be they atoms, molecules or ions – are held in a rigid framework with forces that lock them in position. That's why they are solid. Quickly leaping out of that circular argument, there are also forces acting between the atoms or whatever in liquids and gases. In a pure gas, the particles are whizzing around very fast, and any forces that act between them are too weak to bring that motion to a halt. In a liquid, the atoms are moving more slowly and will become connected to each other for a short while, before breaking free and connecting to other atoms. That is why a solid is rigid and unchanging, while gases and liquids are fluids that are able to flow into any shape. Future work would show that there is a whole suite of forces that can act between the atoms in a fluid. In a liquid these include the forces that hold crystals and other solids together, so when the temperature of the liquid falls to a critical point, these forces spread through the substance, bonding it into a solid.

The weakest forces at play have been named Van der Waals forces after the man who first showed that they existed (although he did not know how they acted). Van der Waals forces emanate from every atom. While we often regard an atom as an unchanging, solid building block of nature, it is really a shimmering mass of randomness. The atom is made from a set of subatomic particles, some of which are charged. The subatomic particles – namely electrons, protons and neutrons – work together to create a neutral blob. However, rather than remain in total lockstep with the other particles, the electrons seethe around the perimeter of the atom. They sometimes cluster at random on one side or the other for a tiny fraction of an instant before dispersing, only to mass together by chance somewhere else. The clusters do not add up to much, but this activity creates tiny electrical forces that push and pull other atoms nearby. It is these minute Van der Waals forces that are ultimately responsible for hauling in those dizzy gas particles and bringing them together into a liquid state.

The so-called permanent gases had so far resisted being liquefied because squeezing their particles together at immense pressure was not enough to allow the Van der Waals forces to act. The only way to liquefy these gases was to cool them so much that their particles slowed to below a critical speed.

Van der Waals forces underlie the cooling effect when a gas expands in your fridge. As the gas particles part company, they need energy to overcome the tiny cumulative forces pulling them back together. They end up stealing that from the surroundings, creating a reduction in overall temperature – and here we are back on refrigeration.

Not only was this new understanding used to bring mechanical refrigeration under proper control, but it also presented a mechanism for reducing temperatures to unimaginable depths, perhaps even to absolute zero itself.

❅

In 1877, it all happened at once when two people figured out how to refrigerate oxygen into a liquid. The most celebrated – and the one who celebrated the most – was Frenchman Louis Paul Cailletet. His technique was to use the Joule-Thompson effect twice, first to chill a vial of compressed oxygen (squeezed to the equivalent of 300 atmospheres) down to −29°C. He then let the compressed oxygen expand through a nozzle, whereupon it erupted as a cloud of droplets that gathered on the side of the collection vessel. Cailletet guessed that the temperature of this liquid – liquid oxygen no less – was about -200°C.

It was Sunday 2 December, and Cailletet hastily dashed off a note for presentation to the Académie de sciences and lodged it with a friend. Cailletet was lined up to be elected to the Académie on 17 December, so he decided to make his elevation more memorable by keeping his achievement secret until the next weekly meeting after that on Christmas Eve.

After a fair degree of pomp, Cailletet's announcement was finally made. Cailletet's dream of scientific greatness had come true, but the reverie lasted for only a few minutes. The academy secretary then piped up that he'd received a telegram two days before from Raoul Pictet, a physicist based in Geneva. It read: 'Oxygen liquefied today under 320 atmospheres and 140 degrees of cold by combined use of sulfurous and carbonic acid.'

We can imagine Cailletet struck dumb as the secretary turned to a letter Pictet had prepared earlier that set out his method, a complex cascade of evaporations, each one cooler than the last. By now Cailletet was protesting his priority over the discovery, having to explain that he'd waited 22 days to go public. Fortunately for him, the letter written on the day of his breakthrough had been dated and sealed and that was evidence enough that he, Cailletet, not Pictet was the first to see droplets of liquid oxygen.

Cailletet had learned his lesson and did not waste any more time. When the Académie met the following week on New Year's Eve, Cailletet was able to announce that he had also liquified nitrogen, which has a slightly lower boiling point than oxygen. His position in history was secure.

Both the methods of Pictet and Cailletet were able to produce only tiny quantities of liquid. What was needed was an industrial process, and that was exactly what Carle Linde had been working on.

We've met Carle Linde before. It was his work on ammonia refrigerants that revolutionised industrial refrigeration in the early 1870s. Not content with that, Linde was already well on the way to producing a system for liquefying air by the time Pictet and Cailletet were making their announcements.

A forward-thinking kind of chap, Linde was looking for a way of using refrigeration to isolate pure oxygen and nitrogen straight from the air. Air is 78 per cent nitrogen and 21 per cent oxygen, with the remaining 1 per cent being mostly argon, with a bit of carbon dioxide, plus tiny amounts of neon and similar gases. Each one of these gases will boil out of liquid air at a specific temperature, and so can be collected separately. It's like distilling alcohol but by using cold not heat.

In 1895 Linde shared the patent for the process with William Hampson, who had independently figured out much the same system over in England.

The so-called Hampson-Linde cycle uses regenerative cooling, which means that instead of cooling a food compartment or freezing ice, the cold that it produces is used to further chill the same supply of refrigerant. The refrigerant used is the air itself, which starts off by being compressed with a pump. This adds energy to the system and makes the air warm up. Next, the compressed air is run through a radiator, which allows the gases to shed some of that heat received as they were compressed. From there the air travels through a heat exchanger, or recuperator, which is simply an inlet pipe wound tightly around an outlet pipe. We'll come back to that soon, but on the first run through it does not do very much.

At the far end of the recuperator, the compressed air arrives at a valve and is squirted through into an expansion chamber. As in an everyday refrigerator, that results in a temperature drop. However, while a modern fridge runs at a pressure equivalent to around five atmospheres, the air liquidators run at pressures at least ten times higher. Air is not an ideal refrigerant, so higher pressures are needed to ensure a large temperature drop as the air expands.

The cold air continues on, entering the outlet pipe that feeds the recuperator. This cold expanded air chills the supply of compressed air that is now making its way through the inlet pipe towards the expansion valve. The cold expanded air then returns to the compressor and is repressurised, warming up a little as before. However, overall the gas is around 10–20°C colder than when it was here the first time. Every cycle brings the temperature of the gas down until it gets so cold that it begins to liquefy in the expansion chamber. That liquid air no longer takes part in the cycle and gradually builds up as more air is cooled. A valve on the compressor lets fresh air into the system to compensate for the loss of gas flowing around as it liquefies.

The supply of liquid air is then tapped off and allowed to warm up again. This process is called rectification and it is at this point that the different constituent gases in the air are collected.

It took Linde until the early 1900s to make the whole process work commercially, and it has remained to this day the main method of isolating pure oxygen (and the other rarer gases in air) and for making liquid nitrogen. Linde's company is still one of the world's leading gas producers.

While the industrial problems were being resolved, scientists untroubled by the profit motive leapt on the process as a chance to push temperatures lower than had ever been possible. As the cycle repeats over and over, the refrigerant just keeps getting colder and colder. How low could it go?

Leading the quest was James Dewar, a hugely ambitious Scottish researcher, who had set his sights on scaling Mount Hydrogen using the Hampson-Linde cycle. Hydrogen was the simplest, lightest, most primeval gas. If it could be liquefied, Dewar felt sure great things would result – for science, and for him.

Dewar had a Hampson-Linde liquidator built at the Royal Institution in London. He added an invention of his own, the vacuum flask, also known as the dewar flask in his honour. Dewar had made the first one in 1892. It was really two flasks, one inside the other with the inner one fused to the neck of the outer one. That created a sealed void between the two, which was evacuated using a vacuum pump. The vacuum jacket created a near-perfect thermal insulator. Whatever was put in the flask, hot or cold, would stay at a constant temperature. A dewar flask might be used to keep coffee warm on the way to work, or to store liquid nitrogen at -200°C. It worked just as well for both.

Dewar's plan was to use his apparatus to liquefy a series of gases in turn. The first would liquefy at a relatively high temperature, and be used as a coolant (held in a dewar flask) to help liquefy the next one on the chain. Eventually, he would create a liquid so cold that it could be used to liquefy hydrogen. This last stage would follow the method used by Cailletet. Hydrogen would be compressed to an incredible 180 atmospheres then chilled with liquid nitrogen. Then the compressed gas would be depressurised into a collection chamber. The resulting temperature drop would bring it to below -252°C, the condensing point of hydrogen.

However, Dewar had a rival in the race, a Dutchman called Heike Kamerlingh Onnes. While Dewar likened his pursuit to a race to the top of the tallest peak, Onnes used another geographic metaphor: 'The arctic regions in physics incite the experimenter as the extreme north and south incite the discoverer.' He saw the field of cold temperature physics as the realm of Dutch science. Dutch explorers had made pioneering forays into the Arctic, van Marum was the first to liquefy a gas, and he and his team were going to make the next big step.

While Onnes proceeded with a calm assuredness, Dewar went hell for leather, driving his equipment to its limits – and often beyond. Several of his assistants were injured in accidents, as the brittle iron fractured in the great cold and high-pressure gases erupted with explosive force. Robert Lennox, his most loyal technician, lost an eye in one most violent accident. Dewar was happy to take the risks with himself and others if it meant that he won the race, and in 1898 he led his battered team across the finish line. He had made liquid hydrogen. The final one of Faraday's permanent gases had been conquered with brute forces, risky behaviour and a dose of ingenuity.

Like all fellows of the Royal Institution, Dewar showed off his liquid hydrogen in public displays. He proved just how cold it was by dunking in a tube of liquid oxygen. That froze solid. He dropped a cork into the liquid, and instead of floating, the lightweight cork plummeted to the bottom like a lump of lead. Even as a liquid, the hydrogen's density was still very low. Yet the liquid hydrogen's chemistry was unchanged, and the tiniest drop would erupt in characteristic yellow flames when set alight.

Dewar had made the coldest liquid in the history of the world. And the following year he went one better, freezing hydrogen at around −259°C, just a few degrees above absolute zero. Dewar felt convinced that these white flakes of ice were the coldest substances possible. And he would go down in history as the man who got there first.

But it was not be be. The finish line moved.

Next door to Dewar's lab in London, two other scientists had been working on a new class of hidden elements now known as the noble gases. John William Strutt, better known as Lord Rayleigh, and William Ramsay had so far discovered argon, krypton, neon, xenon and helium. The gases were noble, because they did not mix with the common ones. They did not react with anything, and as a result had gone unnoticed until now. Argon is actually the third most common constituent of the air, but at less than 1 per cent it was easily overlooked. Helium was first identified in 1868 by telltale colours of light coming from the Sun, but in 1895

Ramsay had managed to isolate the gas in tiny quantities by working with radioactive rocks. By 1899 it was apparent that helium, although four times heavier than hydrogen, had a boiling point of -269°C – and its melting point was less than a degree above absolute zero.

It looked as though Dewar had not plumbed the depths of cold after all. But he was ready for the new challenge, if only he could get his hands on some helium. Helium was much harder to come by than hydrogen, and Ramsey and Rayleigh were not inclined to give the self-important Dewar their hard-won supplies.

Dewar's strung-out, ego-powered laboratory set-up couldn't take the strain. Valuable helium was lost, arguments were common and progress was slow. The work eventually stalled completely. Over in Leiden, the home of Onnes, the slow systematic approach began to pay dividends. New sources of helium were found in oil wells and mines, and Onnes's team of instrument makers, known as the 'blue boys' due to their matching overalls, began to construct the machinery that would liquefy the new gas. It took until 1908 before they were ready. The work began early on the morning of 10 July. After several hours of intense activity, coaxing the machine through its paces, the process appeared to stop. The temperature stuck. Onnes and his team were concerned. A colleague came in to help and suggested that perhaps the reason for the stoppage was that Onnes had succeeded without knowing it.

A light was used to illuminate the collection vessel. There in the electric glare was a tiny sample of liquid helium. The Universe had been made a little colder – even outer space was warmer than Onnes's helium sample.

Dewar had expected that his dauntless taming of hydrogen would lead to great plaudits, perhaps even the Nobel Prize. (though that was always a long-shot, following a patent dispute he had had in the 1890s with Alfred Nobel over the invention of explosives). In the end it was Onnes not Dewar who won the ultimate accolade in physics. But as ever, the next big discovery was just around the corner.

*

As is the way with science, the ability to liquify hydrogen and helium opened up new pathways for the experimenter to follow. Ramsey used liquid hydrogen as a coolant in the race to isolate the first samples of the rarer noble gases, which are present in tiny quantities in the air. And once he had perfected the art of liquefying helium, Onnes used it to cool a host of other materials to see how they responded to extreme cold on the doorstep of absolute zero.

He began with platinum, then moved on to gold, and by 1911 mercury had reached the top of the list. Mercury is a very weird metal, being liquid at room temperature. At -39°C it freezes, and becomes a shiny silver-white solid that looks like any number of other metals. Onnes wanted to know what happened to the electromagnetic properties of the metal when it got colder. It was known that magnetic fields had more of an effect in colder conditions, and that effect was not just confined to metals. Even oxygen exhibited a form of magnetism, and that was most pronounced when it was liquefied. What about electrical conductivity, the ability for a substance to carry electric currents?

The opposite measure of conductivity is resistance, which is the ability of a substance to block the passage of a current. Metals have a lower resistance than non-metals like sulphur, glass or plastics. Mercury has quite a high resistance compared to other metals, but Onnes found that once mercury was cooled to below -229°C, its resistance began to drop. At -267°C the resistance disappeared completely.

Onnes came up with a name for this phenomena: superconductivity.

Electrical resistance is what makes electrified components get hot. The conductor is not allowing the current to travel through unhindered. Its atoms and molecules are getting in the way to some degree. As a result some of the energy that is being carried by the electrical current is getting lost, leaking out in the form of heat. A superconductor has zero resistance so the electrical current can travel through it without any loss of energy.

In the 1980s so-called high-temperature superconductors were developed from materials made of intricate crystals not a million miles away from ceramics. They are only 'high temperature' in the context of the extreme cold in which superconductivity was first seen. If something is a superconductor at above 90K or -183°C, it is working at high temperature. Although far from balmy, this temperature falls within the working range of liquid nitrogen, which is a relatively cheap and safe coolant for these materials. We've all seen the videos where a superconductor levitates above a magnet as if by magic or even whizzes around a circular track unhindered by wheels. Superconductors were certainly science fiction made fact in 1911, and they are still seen as such today. However, we use them a lot. Your local hospital has them in its imaging machines, and universities obviously have plenty, while if you have caught the train downtown from Shanghai's main airport, you will have floated all the way there on board a superconducting hover train.

However, Onnes and everyone else were completely stumped by what they had discovered, let alone having any idea of what to do with it. During his mercury experiment, Onnes had taken liquid helium down to 0.9K, and at this temperature the substance would have taken on another fantastical property. It was a superfluid.

Superfluidity was finally observed in 1937 by research teams working independently in Russia and Britain. While a superconductor has zero resistance, a superfluid has zero viscosity. Viscosity is a measure of how hard it is to alter the flow of a liquid – or gas. Treacle is the go-to example of a highly viscous liquid. It flows but it takes its time. Water has a low viscosity. It plunges and splashes with the best of them. But even water is easy to contain. Just put it in a bucket; it won't go anywhere. That's not true of a superfluid. Liquid helium below about 2K launches an escape from its container. It appears to defy gravity by creeping up the sides and trickling over the rim. In 1938, John Allen, one of the discoverers of superfluids, succeeded in making a perpetual fountain. The superfluid surged out of a nozzle into the air and fell back again, only to go back around and spurt out once more. There was no need for a pump or motor. It's a perpetual motion machine of sorts if it weren't for the fact that it required the world's most advanced and expensive refrigerator to run.

By the mid-1930s, physics was beginning to provide explanations for these super-cold super phenomena. This was the age of quantum physics, which had begun to describe nature in terms of a mind-boggling reality far too small for us to observe in any traditional sense. Down at the quantum level, you can't regard particles as tiny balls pinging around. Instead they are wave-like packets where everything is left to chance. It is not possible to measure all of a particle's characteristics at once. Instead, they have a probability of being one thing or another. It could go either way, and as the particles interact to create the macro-scale world that we live in, it is impossible to predict with any certainty what they will do. Science had always relied on causes having effects. Quantum physics drew a veil over that linkage, which people are still trying to peek through.

Nevertheless, deep cold had offered a chance to observe the quantum realm at the human scale. Superconductivity and superfluidity are quantum effects, where the atoms of mercury, helium or whatever else stop being discrete units. Instead, at the every edge of absolute zero, matter begins to operate like quantum waveforms, intermingling to produce those startling phenomena. And quantum physics now offered a new goal for deep cold, the intriguingly named Bose-Einstein condensate.

❁

It is perhaps about time that Albert Einstein appeared. We know him best for the way he altered our view of space and time and for explaining how gravity really works on the largest scales. However, he had other interests too. One of them might have made him rich: he invented a new type of refrigerator. In cahoots with the Hungarian Leó Szilárd, Einstein patented his design in 1926, promising a chiller that was super-efficient, completely silent and, most importantly, completely safe (early refrigerators occasionally leaked toxic fumes). Szilárd was by no means a spare wheel to Einstein's driving force. It was he who figured out how nuclear fission could run in chain reactions – and it was he who got Einstein to lobby Franklin D. Roosevelt in 1939 about building nuclear weapons. That eventually led to the Manhattan Project, the defeat of Japan, and the Cold War. Whichever way history judges them, never has a pair of novice refrigerator salesmen had such an impact.

The Einstein–Szilárd refrigerator is pretty complicated – too complicated, as it turned, out to compete with the more straightforward compression models, and thankfully Einstein had not given up his day job.

A couple of years earlier, in 1924, Einstein had received a paper from an Indian physicist called Satyendra Nath Bose. The paper set out a way of describing photons (the particles that carry light and other forms of radiation). Einstein was impressed and used his clout to get Bose's paper published. He then extended the work to encompass a broader group of quantum particles. This family of particles became known as the bosons, named after Bose. You'll have heard of at least one boson – the Higgs. Bosons are the things that do the pushing and pulling of the Universe. They mediate the fundamental forces that hold everything together (and keep them apart): the photon does electricity, light and magnetism, the gluon glues the atomic nucleus together, and W and Z particles are at work during radioactive decay (it is suggested that gravity uses a boson called the graviton, only no one can find it; they are literally everywhere, so isolating one is a tricky proposition).

Einstein also figured out that certain atoms, such as the main type of helium, called helium-4, could also act like bosons, if only they could be made cold enough. To achieve that, scientists would have to cool them down to 170 billionths of a degree above absolute zero. But if they managed that they would produce an entirely novel state of matter – the Bose-Einstein condensate, neither solid, liquid, gas or plasma, and never before existing in the history of the Universe. And no one really knew what it could do.

Making the condensate proved quite a tall order, and it would take another 70 years to pull it off. A lot of things needed to be invented in the interim. Perhaps the most important was the development of the laser. While natural light is a jumble of colours all tangled up together, a laser beam can be made to contain just a single colour, or wavelength, of light. In a galaxy far, far away, in the late 1970s and early 80s, Luke Skywalker, Han Solo and the rest of the *Star Wars* gang, were busy using lasers to blast smoking holes in things. This became our understanding. Lasers made things hot. However, at around the same time, an American researcher called Steven Chu was figuring out a way of making the atoms in a gas cool with lasers.

Atoms absorb and release photons – in this context we are back to calling these particles of light – and each type of atom will only handle photons of specific wavelengths. So by tuning a laser to the correct wavelength, a photon from a laser beam can hit an atom and be absorbed and then released. If the atom is moving towards the photon as they collide, the laser fools the atom into releasing more energy than it took in. The result is that the atom slows down. In other words it gets colder.

A team at the Massachusetts Institute of Technology (MIT) began to use this technique to cool gases to the lowest temperatures yet. The conventional mechanical coolers could not get any nearer to absolute zero. The closer they got the more effort was needed to make the next step. It is impossible to actually reach absolute zero. It would take an infinite amount of time to extract ever smaller amounts of heat. You'd never get there, but laser cooling offered a chance to get very close.

Laser cooling is a completely haphazard affair. While a few of the atoms are slowed to a cooler temperature, others are unaffected by it. So a second piece of equipment was needed. This was the magnetic trap, a magnetic field that surrounded the gas sample like a bowl. The trap allowed for a form of evaporative cooling, albeit one that was light years away from the methods used by those Egyptian slaves working on the roof night after night 4,000 years ago.

Laser cooling and magnetic traps finally opened up the possibility of making a Bose-Einstein condensate. The first attempts were led by MIT's Daniel Kleppner, a pioneer of laser research. He opted to use hydrogen as his raw ingredient. Progress was necessarily slow. It took years to build and perfect the different machines needed. Just as with Dewar and Onnes, other competitors joined the race.

Not all atoms can form Bose-Einstein condensates. Helium-4 can and so can hydrogen, and there are a large number of others that could do it as well, at least in theory. The prevailing wisdom was that the lighter atoms would be easier to work with. Eric Cornell and Carl Wieman from the University of Colorado, Boulder, decided to give the process a try with rubidium, a metal with atoms 55 times heavier than hydrogen.

In 1995, the pair succeeded. They had vapourised the rubidium, and the gas was held in the magnetic trap. The atoms that were cooled by the laser did not have the energy to bounce out of the magnetic 'bowl'. Those that did evaporated away, leaving the cooler ones behind. As the sample got cooler it also diminished in size. The question was whether anything would be left by the time the temperature reached the critical level.

The magnetic trap was gradually shrunk until just the coldest atoms lurked at the very bottom. By the time the rubidium reached the critical temperature of 170 nanokelvin (170 billionths of a degree above absolute zero), there were just around 2,000 atoms left. But that was enough. The first Bose-Einstein condensate in the history of the Universe formed in Colorado.

As they condensed, the rubidium atoms lost their distinct identity and began to behave as a single mass of quantum material that scientists are still trying to understand.

So what next? It is likely that the Bose-Einstein condensates will find their way into the machines of the future, but we have already had our world changed by the science of deep cold. Forget food – cold technology is at work in the most unlikely of places.

ELEVEN

The Hidden Chill

You can converge a toaster and a refrigerator, but those things are probably not going to be pleasing to the user.

Tim Cook, 2012

When is a refrigerator not a refrigerator? The obvious answer is when it is an air conditioner, but as punchlines go that's not all that inspiring. A refrigerator can also be a gas factory, a rocket engine, a server farm and even a fusion bomb. It's used to dig holes, make dams, track subatomic particles, image the brain and feed half the world (without chilling food, either). This is the hidden chill. Quietly and without fanfare, cooling technology runs deep within the workings of modern civilisation.

In a US home, refrigeration is sucking up a third of the total energy consumption – but most of that is for chilling the air, not the food. At first glance it might seem that the air con is pumping out hot air from inside the house and sucking in cool air from the outside. It's a little more complex than that, and much more effective. In essence air conditioning works just like a fridge. It has a compressor, expansion valve and two heat exchangers. The difference is that the material being cooled is not a food compartment but a flow of air drawn from the body of the house. This air is pulled over the refrigeration coil and chilled accordingly before being pumped back into the rooms. The radiator coil, the hot end of the machine, is connected to the outside – it is often on the outer wall – and it warms a flow of external air to shed the heat extracted from inside.

This kind of air conditioner was invented more or less by chance. On a stifling summer's evening in 1902, Willis

Carrier, a young engineer, was called in to solve a problem at a Brooklyn print works. The ink would not dry, at least not fast enough – the air inside was just too humid. Carrier's solution was to build a machine to dry the print work's sultry atmosphere, so the fresh print runs would stop smearing so badly. He did this by chilling the air with a refrigeration system, so the fog of humidity condensed into water. The dehumidifier was by no means small, but it did the job. It also chilled the air, and the resulting coolness was welcomed by the workers who tended the giant presses. Instead of being a clammy hellhole, the print works had become a rather pleasant place to be on a hot summer's day. Within 20 years Carrier's chillers were becoming common in other industrial settings. It would not be long before everyone wanted some A/C.

The use of domestic air conditioners went hand in hand with the development of mechanical refrigeration. The paired technologies first found their calling in the subtropical United States, the so-called Sun Belt that runs from the deserts of southern California to the swamplands of Florida. Populations have been on the rise in this torrid swathe since the 1960s, mostly because of the availability of inexpensive air conditioning, which transformed the quality of life there from merely tolerable to positively desirable.

Air conditioning is also an essential feature of that other American innovation, the skyscraper. It is often said that a skyscraper's windows are sealed shut to prevent people throwing themselves out of it after some kind of mishap. The real reason is that opening the windows would disrupt the building-wide climate control. Most of a skyscraper's floor space is some way from the glass exterior and would receive none of the cooling breeze an open window could supply. Instead all the air flow in the building is managed by the HVAC (heating, ventilation and air conditioning) system located on the mechanical floor, normally in the basement. If a lot of windows were opened on the sunny side of the building, they would draw in warm air and make many of the spaces deeper inside intolerably hot.

All in all, two-thirds of the buildings in the United States – from normal suburban homes to those cathedrals of engineering downtown – have air conditioning. All that cooling uses 5 per cent of all the electricity produced by US power plants, adding up to an $11 billion bill at the end of the year.

This may be an overly simplistic notion but one worth making: air conditioning is designed to chill the air at the hottest times of the year – in some places that's all year around. However, in so doing, air conditioning is one of the biggest net producers of atmospheric carbon dioxide, at least through the power it consumes. The average home air conditioning unit results in two tonnes of carbon dioxide being added to the global total every year. The United States alone has 50 million AC units, and it is by no means an all-American problem: Australia, the Persian Gulf, India and China have all got the air-con habit, and many other countries are not far behind. It's perhaps a bit of a careless corollary, but all that air chilling is making the air warmer.

There is another way, one that perhaps should be called a *shabestan* or *badgir* in homage to Persian know-how but is better known as the evaporative cooler. This is a modern take on an old-fashioned method to cool air. The cooler only really works in arid climes, where a dry wind can be harnessed to create a jet of cool air. The evaporative cooler is an open-sided box located on the roof or elsewhere high up on a building. The box is lined with woollen pads, which are kept perpetually soaked with water. This water evaporates when hit by a blast of hot air arriving from outside. That evaporation steals away some of the heat of the air, creating a reservoir of cool, humid air inside the cooler box. A fan then pumps that down into the house, displacing the warmer air contained within.

The cooling mechanism harnesses a natural effect, and the cooler consumes only a small amount of power to drive a small water pump and air blower. Evaporative coolers are certainly cheaper to run than air-con units and are one of a growing squad of greener cooling techniques that are being incorporated in modern building designs.

You can't get much more of a modern building than a data centre, otherwise known as a server farm. These are warehouses full of computing power run by governments and the likes of Google, Facebook, Amazon and the other big technology firms. A server is a behind-the-scenes computer that stores the information that you access from your personal gadget. The 'cloud' may sound better, but really all your online photos and docs are better described as being stored in a 'shed'. These state-of-the-art sheds can get extremely hot. Google alone has nearly a million servers distributed all over the world. Each server sucks up a lot of energy – the big data centres use the same electrical supply as a town of 80,000 people – and the computers pump out a great deal of heat as they serve the world.

In some cases data centres are chilled with immense air conditioners, but as you might expect, the tech giants have got a little bit more innovative when it comes to keeping their cool tech chilled. They do things that might not work too well with a personal computer – they give the servers a bath. There are no whirring fans in the servers, and hard disks are sealed tight. Everything else is flooded with oil, which pumps through, whisking away the heat. Google apparently does the same thing but swap oil with waste water. Meanwhile Facebook have gone natural. Their European data centre is just a few miles below the Arctic Circle in Sweden. The huge energy demand is delivered by dams on the nearby river, while the average temperature up there hovers around freezing so simply letting cold air in from outside will be enough to cool the equipment inside. Basically they open the window.

We are a long way from being able to reduce the polluting effects of air conditioning, and its potential for destruction is not to be sniffed at. However, there was a much more chilling example of the power of chill. Refrigeration was used to make one of the biggest bangs in history: the first H bomb.

The H in H bomb is for hydrogen, but it's not the ordinary stuff ignited by Cavendish or liquefied by Dewar. It is a mixture of heavy isotopes called deuterium and tritium. Ivy Mike was the code-name for the first H bomb test, which took place at Enewetak Atoll in the Pacific on 1 November 1952. This was a nuclear weapon, but as well as employing nuclear fission (splitting of atoms) it also used nuclear fusion to boost its explosive power. Nevertheless, it's not quite a fusion bomb that might be deployed in an episode of *Star Trek*; mercifully that's still science fiction. Instead Ivy Mike was the first thermonuclear weapon, making use of both fission and fusion to utterly obliterate a small corner of the Earth.

It was really three bombs in one. A conventional high explosive went off to create a force that was focused inwards on a warhead of enriched uranium. The pressure started a fission chain-reaction (much like the Fat Man bomb at Nagasaki), and this fission created the great heat and pressure needed to make the atoms of heavy hydrogen fuse – releasing the largest artificial explosion seen up until then on Earth (nuclear fusion is a powerful thing. It's the process that powers a star).

When the bomb was primed, the hydrogen 'fusion fuel' inside was liquified using an immense cryogenic apparatus, which by itself weighed 18 tonnes. The Ivy Mike took weeks to assemble, and therefore was not ideal as a rapid-response weapon. In its place, the US deployed a class of exploding radioactive thermos flasks, known as the Mark 16 nuclear bomb. These were small enough to be carried in a bomber aircraft. The cryogenic refrigerator of Ivy Mike was replaced by dewar flasks filled with liquefied deuterium. If the call came, these would have been used to fill the bomb before it was dropped. By 1954, insanity prevailed and the measly 6-megaton Mark 16s were replaced with 15-megaton Mark 17s. These used a dry fusion fuel made from radioactive lithium, which did not require refrigeration.

The arms race was not the only beneficiary of the ability to liquefy gases on an industrial scale. Mercifully, history relates that instead of wiping out half of the human race, half of the human race now relies on a refrigeration technique to survive.

❄

Fritz Haber, the German chemist, was a controversial figure. He invented a method for extracting nitrogen from the air, which changed the world in more ways than one. It ensured that a human population could boom well into the billions, but it also made it possible to build explosives on an unprecedented scale. The two are not mutually exclusive.

Nitrogen is the most common gas in the air. It is essential for the synthesis of proteins in all living things, but is remarkably difficult to get at. The natural world 'fixes' the gas from the air through a complex cycle that involves weird primeval bacteria in the soil. These bugs make nitrogen compounds available to plants – and then on to us and to the rest of the natural world through plant foods.

Natural fertilisers, such as dung, night soil, guano – whatever you call it, it's basically the same stuff – are full of the all-important nitrogen, which is why they help crops to grow.

By the 1900s, William Crookes, a British physicist better known for inventing cathode-ray tubes, predicted that humanity was facing a Malthusian future. Soon the global population would grow beyond its ability to harvest food. What was needed was an artificial source of nitrogen to fertilise more crops than was possible with the natural cycle of nutrients. There were geological sources of nitrogen compounds that could be mined, but what was needed was an industrial process to 'fix' a supply of nitrogen from the air. By 1908 Fritz Haber had developed one that worked. It took some refinement by others but we generally call it the Haber process today.

The Haber process mixed nitrogen and hydrogen gas in such a way that it produced large amounts of inexpensive liquid ammonia. It's complicated stuff involving catalysts, and the actual reaction takes place at high temperatures and pressures. However, the cleverest bit was removing the ammonia from the system once it was produced. If it were left in too long it fell apart again into the raw ingredients. Although it does not use mechanical refrigeration as such, the Haber process makes use of a kind of recuperator, or regenerative cooler, like the one in a Hampson–Linde air liquefier.

Haber and his colleagues found that they had to scrub the raw ingredient gases clean of impurities before they reacted. One cleaning process involves pumping the gases through a spray of water, which cools them down. Rather ingeniously, these cool gases are then used to chill the hot reacting mixture further down the line. That makes the ammonia being produced inside condense into a manageable liquid that can be removed, and that in turn allows the reaction to continue at full tilt.

More than 130 million tonnes of ammonia are made each year by the Haber process. Some of it is used as the refrigerant in industrial cold storage or the HVAC system in your local skyscraper. However, 85 per cent of all ammonia is used to make fertilisers. In the 1960s, the UN-backed Green Revolution spread Haber's technology around the world, making famine a thing of the past in many developing nations. It is estimated that at least three billion people eat food that only grows because of Haber's technique for making fertiliser from thin air. We might assume that those three billion live in the developing world, but developed nations have simply been relying on fertilisers for longer. So every two days or so we all take our turn as one of those three billion.

Since he saved the lives of half the world's future population, it's an understatement to say that Haber's process had an impact on civilisation long-term. Nevertheless it has had some short-term impact as well. His first plant came on stream in 1911. Without its ammonia, the First World War would have lasted only months as Germany would have run out of natural supplies of nitrate used to make explosives (the British controlled the main sources in South America). Haber was also instrumental in the development of the first deadly chemical weapon, the chlorine gas released at Ypres in 1915. So, as we've said, he certainly left his mark on the world.

❅

The hydrogen used in the Haber process is derived from natural gas, otherwise known as methane. Route One for methane is the gas pipe, an intercontinental tube that blows this useful gas to where it is needed with the aid of a jet engine or two. However, pipes generally have to stop at any large sea or ocean, so in the past half-century or so a new transoceanic route has been developed – a floating refrigerator otherwise known as an LNG (liquefied natural gas) carrier. LNG carriers are the new kids in town, sidling past the oil tankers on the high seas as they deliver gas to customers beyond the reach of pipelines. Tankers outnumber carriers by around eight to one, but the ratio appears to be on the slide.

Natural gas is transported as liquefied NG instead of plain old NG because, once cooled to −162°C, the liquid takes up around a 600th of the volume of the gas. The LNG is cooled on land and pumped into vast insulated tanks for the journey – the actual process is a little involved to stop the tanks buckling under the shock of receiving a huge surge of cold liquid. The carrier's engines power compressors which keep the LNG under the correct pressure for the duration.

The ships may be at sea for weeks, labouring through the warmest parts of the oceans, and even the thickest insulation is not enough by itself to keep the tanks cool. Therefore LNG carriers use a phenomenon called auto-refrigeration. The tanks are kept at the exact pressure required to hold the liquid at its boiling point. The surface of LNG is constantly evaporating into NG, and that change of state has its own cooling effect. As long as the LNG is kept at a very gentle simmer during the voyage, its tanks will cool themselves.

Once arrived, LNG is generally converted back into gas, so it can enter the local pipeline grid, but the liquid form is also available as pumped fuel for vehicles. LNG fuel is regarded as being cleaner than gasoline or diesel. It produces fewer nasty pollutants and releases less carbon dioxide, but running a car on liquids cold enough to freeze just about anything in seconds has its challenges.

The other gas used in the Haber process, nitrogen, is isolated by simply burning some hydrogen in air. That combines with all the oxygen mixed in to make water vapour, which is condensed away and leaves more or less pure nitrogen behind. Purifying oxygen is a lot harder. It has to be liquefied. As previously discussed the air is cooled to -200°C using air liquefiers developed by Carl Linde and William Hampson. At this temperature the air becomes a liquid. When left to warm up slightly, the nitrogen in it boils away (at around -195°C), and can be collected as a pure gas. Next comes argon (-185°C) and that leaves pure liquid oxygen, which boils into a gas at -183°C.

In case anyone is wondering, the small amounts of carbon dioxide and water vapour in the air will have frozen during the initial cooling process and have been removed already, so the gases produced are completely dry.

Pure nitrogen gas finds its way into bags of washed salad, where it impedes the decay of the delicate leaves. If too much oxygen gets in, the leaves will go brown and slimy. It is also the gas of choice for filling the tyres in aircraft and racing cars. Aircraft wheels have to stay adequately pressurised even during the cold conditions of cruising altitudes. Any water vapour in the tyre would freeze in flight and weaken the tyre. Nitrogen is used in Formula 1 and Champ Car tyres for the opposite reason – it does not get too hot while the cars accelerate and decelerate around the track.

Oxygen, an altogether more reactive substance, has a more gritty purpose. Half of all the pure oxygen produced by liquefaction is used to make steel. A jet of gas is blasted through the molten mixture of iron and slag to burn away those extra impurities. Without this 'oxygen lance' process, the toughened steels that support the modern city, span the world's rivers and are welded into the biggest ships would be impossibly expensive to produce on any reasonable scale. All those things are only possible because of oxygen coming out of a fridge – but there is more.

A quarter of the world's supply of liquefied oxygen is used to make plastics and other artificial fabrics (many of which have to be produced in low-temperature environments, thanks also to refrigeration).

The air tanks used by divers, doctors and firemen are little more than dewar flasks filled with precision mixtures of oxygen and other liquefied gases. And most of the rest of the oxygen is used for what oxygen does best. It makes things burn. The super-hot flame of an oxyacetylene cutter was born in a refrigerator, as was the fuel that launches the world's heavy-lifting rockets. All space-rated liquid-fuelled rockets, from the V2 space bombs of the Second World War to the Saturn V moon rockets, make use of liquefied oxygen to burn their fuel. The most powerful, such as the retired Space Shuttle and the Delta IV Heavy that is being tested for a putative Mars mission, are also filled with liquid hydrogen.

The liquid fuels are sprayed together inside a combustion chamber to create the explosion of thrust that pushes the payload into orbit. It might not look like it but a liquid-fuelled rocket on the launchpad is a giant refrigerated vehicle. The fuels have to be cooled to a liquid so that the engine can combine enough of them together sufficiently fast to create the required bang. Puffs of gas would just not create enough thrust.

Filling a tank with liquid hydrogen rocket fuel is extremely difficult. The techies call the problem BLEVE – a boiling liquid expanding vapour explosion, which has a nice ring. Generally the trouble involves the three Es: evaporation, expansion and explosion, which are never good and all require different approaches on the surface, in the air and in space.

Liquid hydrogen must be kept at below –253°C to prevent it from boiling back into a gas, which would create intolerable pressures inside the rocket. The fuel tanks are heavily insulated, not least from the heat of the rocket engine as it fires, but also from the direct heat of sunlight out in space. There is always some hydrogen gas boiling off, and the tanks must be able to vent this safely away, without risking an explosion. To complicate the issue further, liquid hydrogen can leak through welded metal seams, so a cryogenic rocket-fuel tank has to be 'inflated' with liquid hydrogen, a bit like a balloon. The seams between the sections of the corrugated walls of the steel tank (all only millimetres thick) are squeezed tightly by the huge internal pressure of cold liquid, thus preventing leaks.

Once the rocket engine fires, its liquid fuels go from around -253°C to 3,315°C in the blink of an eye. All that release of heat is what is creating the thrust of exhaust gases from the back of the engine. A regenerative cooling system employing the cold fuel itself is used to mitigate the extreme temperature changes that would otherwise risk burning away the whole engine. Before they are allowed to meet inside the engine's combustion chamber, the cold fuels are pumped through a jacket that surrounds much of the rocket engine, including the familiar nozzle at the back. This stops the engine components from getting overly hot, while at the same time it warms up the fuels before they burn away at the heart of the rocket.

Liquefying air yields more than oxygen and nitrogen. As mentioned, argon is released from liquid air, while krypton and xenon, two more noble gases present in minute amounts in the air, have also liquefied on the way down.

The air is also the primary source of neon and helium. There is not much of these gases there, just eight and five atoms for every million respectively, but both are very useful so it is worth taking the air down to the deepest of temperatures to collect them.

Neon is obviously most famous for its use in the neon lights which lit up cities for much of the twentieth century. Fittingly it was Paris, the City of Light, that saw the first large-scale use of neon lighting. Its chief advocate was French industrialist Georges Claude, who developed neon lights in 1910 as a way of selling the byproduct noble gases (otherwise useless at the time) from his air-liquefaction business, which was aptly named Air Liquide.

The lights are examples of gas-discharge lamps, in which a diffuse supply of gas held inside a sealed tube is made to glow with a characteristic colour by an electric current. Actually lights filled with neon only glow red. The other colours of 'neon' lighting are produced by the other gases.

The noble gases are generally harmless when breathed in. The exception is argon, which has a rather debilitating effect. Xenon is used as a breathable anaesthetic in rare cases, while helium has many more applications, not least as part of a breathing gas used in medicine and by divers. Replacing at least some of the nitrogen in air with helium makes a mixture that is easier to breathe by patients with respiratory problems, and is also safer for divers going to depths below about 40 metres (130 ft).

A similar mix of gases is used to fill children's party balloons. A substance that must be harvested close to the lowest temperatures imaginable ends up in an inflatable cartoon character that delights and bamboozles toddlers the world over. Some would say that's a good use of helium, but it leads to dismay in more cerebral quarters. In 2012, Professor Tom Welton from London's Imperial College pointed out that we are running out of helium, and will suffer severe shortages within 50 years. He told the BBC 'The reason we can do MRI is we have very large, very cold magnets – and the reason we can have those is we have helium cooling them down. You're not going into an MRI scanner because you've got a sore toe – this is important stuff. When you see that we're literally just letting it float into the air, and then out into space inside those helium balloons, it's just hugely frustrating. It is absolutely the wrong use of helium.'

Liquefying gases is not just a means of purifying them. They are used as the world's best coolants too. As well as being used in MRI scanners, liquified gases have led to test-tube babies, life-saving drugs, and the largest scientific experiment in history.

*

If you've ever had the need to lie inside an MRI scanner, you will have become a living magnet and a living radio transmitter – for a few fractions of a second anyway. The MRI or magnetic resonance imaging system sends an immense magnetic field through your body. That makes the molecules in the target area being scanned line up a little like compass needles all pointing north. When the magnetism fades away, the molecules flop out of alignment again and emit tiny broadcasts of radio waves as they do so. It is these radio waves that are picked up and used to create an image of your bits and bobs.

A magnetic field powerful enough to do this is made by a superconductor, and that needs to be cold to work: less than 10K (-263°C). Liquid helium does the trick, taking it down to about 4K. The supercooled magnets are bathed in liquid helium inside a doughnut-shaped cryostat – basically a clever dewar flask. It all has to be sealed up and heavily insulated. That is why the patient has to be inserted right inside the machine, which can be unpleasant for some.

The same class of magnets is used in maglev trains – short for magnetic levitation. The concept for these trains is rather old-fashioned but they remain a vision of how transport can be faster and better. Instead of using wheels spun by a rotating motor to move the train along, a maglev employs something called a linear motor. The train carriage and the track are fitted with superconducting magnets. Like any magnet they have a polarity – one end is a north pole and the other is a south pole. These subscribe to the famous rule of opposites attracting and likes repelling. Carefully targeting repulsive magnetic forces lifts the train off the ground. Then by flipping the polarity of the magnets along the track in sync with those on the train, a wave of magnetic forces can be created that pushes the vehicle along.

Wheeled transport relies on friction between the wheels and the ground. The friction resists the motion of the wheel and therefore the wheel (and its vehicle) move forwards. However, friction here and elsewhere in the moving parts creates heat, and this limits the speed and efficiency of wheeled vehicles. The maglev, hovering above the track like a tube train just arrived from the future, is unencumbered by friction so can whizz along at record speed. The top speed achieved so far is 581km/h (361 mph) which is frankly only a whisker above the record speed for a wheeled train.

A handful of maglevs run commercially. They are quiet and so far reliable, but eye-wateringly expensive to build. Whether the future of transport really lies in trains floating above superconductors cooled to near absolute zero by one of the world's rarest substances remains to be seen.

By contrast, it is clear that the future of particle physics lies in using supercooled superconductors. In 2008 the Large Hadron Collider (LHC) was completed, a particle accelerator buried under the border region between France and Switzerland. Run by CERN, the European research institute, the LHC works like many other particle accelerators, only it is bigger and better. A bit like a maglev track, it uses the force fields of superconducting magnets to control a beam of subatomic particles. Guess which type: yep, large hadrons. Basically, for our purposes we can just think of these as protons, but at a build cost of £6 billion you can rest assured that the thing can handle more than that.

In 2013, the LHC presented evidence for the Higgs boson, a particle that gives matter its mass. Mass is the quality that causes matter to resist being pushed around by forces, and is therefore pretty fundamental to the way everything works. So the discovery of the Higgs is up there with the greatest moments in science. While the electron was found with a glass vacuum tube, the proton showed up using some photographic paper and gold foil, and the theory of relativity was confirmed using photographs of a solar eclipse, the Higgs boson required 27 tonnes of superconducting magnets. There are 1,600 in total arrayed along the 27-kilometre subterranean ring, all cooled to 1.9K (-271.25°C) with 80 tonnes of liquid helium. The LHC is the largest refrigerator that has ever existed.

The Higgs boson was detected using instruments that made use of some mind-boggling electronics. In the earlier days of particle accelerators, the detectors used were called bubble chambers – and they were refrigerators as well.

The bubble chamber was invented by American Donald Glaser in the 1950s, as an improvement to the cloud chamber which was the main particle detector in the early twentieth century. The cloud chamber was invented by the Scot Charles Wilson, an avid hill walker who got the idea while hiking on the misty mountains near Ben Nevis. The story goes that Glaser got his idea while gazing at the bubbles in a cold beer.

A cloud chamber was able to reveal the track of particle as a momentary streak of cloud in water vapour held at dew point (just as it was beginning to condense into droplets). Glaser's idea was to do the same thing but use a tank of cold liquid. The inspiration of beer is probably apocryphal, but Glaser did report trying out his concept with a tank of chilled beer. He found that liquid hydrogen worked a lot better. As the particles are entering the chamber, the pressure is dropped very suddenly making the liquid begin to transform from liquid to gas. In that instant, the particles create a track of bubbles through the liquid which are easier to detect than those in a cloud chamber.

Charles Peyrou, a French physicist, put it like this: 'Cloud chambers are like ladies, very delicate to handle, but bubble chambers are like prostitutes: they will work for anyone.' Unsurprisingly, they built several of these detectors at CERN, including one called the Big European Bubble Chamber. That name perhaps lacks the gravitas of the Large Hadron Collider, but it detected all kinds of exotic particles that paved the way in the hunt for the Higgs.

❄

If you need to make something very cold then helium is the coolant of choice. It is inert and so doesn't not react with anything. However, it's expensive stuff. Liquid nitrogen is 250 times cheaper. It is also more or less inert and if -195°C is cold enough for you, then it will do the trick. Liquid nitrogen is the workhorse of a field of science, medicine and engineering called cryogenics.

Apart from the usual caveats about handling such a cold substance, liquid nitrogen is more or less harmless inasmuch as it won't corrode containers, poison foodstuffs or chemically alter most of whatever goes into it. However, breathing pure nitrogen is something to be avoided, and in 2013 a pool party in Mexico showed just how dangerous liquid nitrogen can be.

The party was held in León as a promotion for a drinks company. The climax of the event involved pouring dozens of litres of liquid nitrogen into the pool. The nitrogen immediately boiled away, and like dry ice, the cold vapour created an impressive cloud effect that billowed across the pool to the delight of guests. However, as the cloud began to clear, it became obvious that something was very wrong. The party organiser had not understood that the stunt would cover the pool in a thick layer of nitrogen – colourless and odourless, but deadly. People in the pool were in various states of asphyxiation. Several had to be rescued from drowning and one of them was so badly brain-damaged that he fell into a coma.

However, liquid nitrogen saves more lives than it takes. It is used to fast freeze foods and medicines being rushed to disaster zones. There are perhaps people reading this and many more who have children who began their lives frozen in liquid nitrogen. The sperm, eggs and embryos used in in-vitro fertilisation and other fertility treatments are stored in liquid nitrogen. The freezing and thawing process carries some risks to the cells, but once safely frozen, there is no limit to how long they can be stored.

Liquid nitrogen and other coolants are used to create cryogenic chemical reactors, where the intricate molecules required for modern medicines can be produced with great precision. One of the drugs produced using cryogenics is the statin Lipitor. Otherwise known as atorvastatin, this is a cholesterol drug taken by millions every day. From 1996 to 2012, $125 billion worth of this drug was sold, making it the world's best-selling medicine of all time. In the UK and USA more than one in eight people pop statin pills to keep their cardiovascular health in check, yet another example of how the application of cold impacts lives in unseen ways.

Cryogenics is used in engineering too. One of the favoured tricks of science demonstrators showing off the cooling power of liquid nitrogen is to dunk a rubber ball in it. The once-squishy elastic ball goes rock hard, and when thrown to the floor it does not bounce back but shatters like a piece of glass. Materials that are soft and flimsy at normal temperatures become tough solids when taken down to cryogenic temperatures. This means that soft plastics can be carved as if they were marble into intricate shapes, or ground into ultra-fine powders. The technique, known as cryomilling, can be done on an amazingly fine scale, down to millionths of a metre. Genetic material has even been carved out of a frozen sample, like a fossil chiselled from the bedrock.

However, the chief use of cryogens, like liquid nitrogen and dry ice, is for something altogether more familiar: powdered soups, instant coffee and noodle snacks. These foods are fully ready to eat, but have had all the water removed by freeze-drying. The application of boiling water reconstitutes them into something near to their original state. They are seldom very tasty but incredibly convenient which is why they are invaluable to soldiers, spacemen and students who don't have the time or means to create food from scratch.

Freeze-drying was developed in the 1940s as a way of preserving medicines and blood products without the need to keep them refrigerated. The process requires a good deal of finesse depending on what is being dried, but basically the item is frozen and then exposed to a vacuum, which makes the ice sublimate away, leaving just a tiny fraction of the original water in the material. Fruit juices are treated in a similar way to create concentrates that are later reconstituted with water into something like the original juice.

The freezing can be done using mechanical refrigeration, but in the case of foods it needs to be done very fast to prevent the ice crystals from destroying the intrinsic structure. As a result, flash freezing using cryogens – perhaps liquid nitrogen but also dry ice – is often the best way to do it.

The freeze-dried products will reabsorb water from the air if left exposed to it. As a result they have to be sealed in an airtight packaging, perhaps flushed with nitrogen gas to ensure that no air remains. Freeze-dried foods are lightweight, compact and last for many months. However, water is not the only component in the food that will sublimate – oils and vinegar will go too, which is why freeze-dried food always tends to have that distinctive taste (taste being a relative term here).

As any mother will tell you, freeze-dried food does not make a substantial meal. Cryogenics is capable of making something much more substantial, only we cannot eat it.

❄

If you were to list the objects that require refrigeration, you'd no doubt come up with a pretty long list. However, few of us would include concrete. We seldom give concrete much thought, but perhaps we should. After all, it is artificial stone, a slurry of liquid and powder that becomes solid as a rock in whatever shape we choose. The science behind concrete is enormously complex. In fact no one is quite sure what exactly happens, but when it is drying it produces heat. In mass concrete structures, such as a dam or the foundations of a large building, these temperature shifts are often uneven and this means that the middle of the structure is expanding while the outside is contracting. That leads to cracking, which can never be a good thing when it comes to concrete buildings.

To avoid such disasters the concrete is cooled in different ways, either with a nozzle of liquid nitrogen that blasts into the concrete mixer like a high-end cocktail gone mad, or by embedding refrigerant tubes inside the concrete as it sets.

Liquid nitrogen has other uses in construction. Digging a shaft into the ground or boring a tunnel is a whole lot easier if the perimeter of soil has been frozen with injections of liquid nitrogen. A similar system is used to create so-called freeze walls around mines. The idea here is that the ice forms a barrier to any nasty pollutants being unleashed inside so they do not escape into the wider environment.

The most famous freeze wall is being built (at the time of writing) around Japan's Fukushima nuclear power plant, which suffered a series of meltdowns after the 2011 tsunami. The 1,500-metres long and 33-metres deep barrier is not the largest of its kind, but it will have to stay in place for decades. The groundwater that floods into the plant everyday cannot be allowed to seep away into the surrounding area, for fear that it will make the water table radioactive. Hence the freeze wall. While the wall is being built, groundwater is being pumped into temporary storage tanks, something to which the ice barrier will hopefully put at an end.

While the future of the Fukushima ice wall – at least its effectiveness – is uncertain, it shows yet again that the power to control cold is one of humanity's greatest abilities. What that power will create in future is a guessing game, but let's have a go.

TWELVE

The Future is Cold

Never let the future disturb you. You will meet it, if you have to, with the same weapons of reason which today arm you against the present.

Marcus Aurelius Antoninus, *c.* 175 AD

Refrigeration has changed the world, so how might it do it all again? Predicting the future is a fool's errand. Let me be your fool for a little longer. Kevin Kelly, a founder of *Wired* magazine, puts it this way: 'Any believable prediction will be wrong. Any correct prediction will be unbelievable.' So let's bear that in mind.

The world needs new fuels and new ways to store power. It's not possible to turn renewables like solar power and wind energy on and off when we need them. They make electricity when the correct conditions prevail, and frequently that won't tally with when we need the power. So the race is on to find ways to store that harnessed energy in a reusable form.

Liquefied hydrogen is often mooted as the fuel of the future. It can be made by the electrical splitting (electrolysis) of water, chilled into a liquid for storage, then burned to release heat when required. The only pollution from a hydrogen-powered engine is water vapour. Imagining a world with liquid-hydrogen filling stations in ordinary places takes a big leap. At the moment it is only extraordinary places – rocket launch pads and the like – that are set up to handle this extremely tricky substance.

However, there are research programmes that are looking for ways of making hydrogen safer to handle, and many of them are inspired by a strange form of natural fuel: an ice that burns.

The permafrost of Siberia and sub-Arctic Canada is a source of natural gas. Over the millennia the organic material in these frozen soils has decayed very slowly and released methane, or natural gas. In the 1970s it was discovered that some – in fact a lot – of the natural gas gets locked away in a form of water ice called a clathrate. The methane is trapped underground at high pressure and low temperatures inside a mush of ice crystals. The clathrate looks like a dirty slush puppy ice drink. There is a big difference: when lit with a match, the ice bursts into flames.

Clathrates are packed with natural gas and offer a cleaner form of fossil fuel, if it can be got at, that is. As well as being found in deep rocks, clathrate deposits have been located on cold sea beds where the temperatures hover around freezing. In 2000, a Canadian trawler hauled a tonne of clathrate ice to the surface in its nets by mistake. The mass of ice hissed menacingly on the surface as the gas escaped at the lower pressure. The trawlermen decided to drop it back into the water and get out of there.

Engineers frequently copy nature, and it has been proposed that clathrate-like ice could be used to hold hydrogen as its 'guest molecule' instead of methane. Hydrogen clathrates might one day be a stable form of hydrogen fuel, pumped as slush into our fuel tanks.

In 2006, researchers went one better. They created a whole new kind of water ice, which could also be used as a fuel store. Wendy Mao, working at Los Alamos National Laboratory in New Mexico, squeezed ice to enormous pressures using a diamond anvil. This is a gadget that is designed to recreate the pressures under which carbon forms into diamonds deep inside the Earth. We are talking in the order of six million atmospheres. Mao and her team then bombarded this compressed ice with X-rays and found that the water molecules in the ice separated into an alloy of hydrogen and oxygen. An alloy is normally used to describe a mixture of metals, where one type of metal is mixed into another, effectively dissolved in it. However, this weird ice, kept solid at high pressures, can also be understood as an alloy of hydrogen molecules mixed with oxygens.

Mao's water alloy is nothing like normal ice. It is brown and it did not melt even at temperatures of 400°C when kept at enormous pressures. Again, here we have a possible future means of transporting and storing clean hydrogen fuels.

As well as being a source of clathrates, ocean water could offer a source of renewable energy, and one that is not so reliant on prevailing conditions. The mechanism being proposed is a steam engine that works like a refrigerator in reverse.

The system makes use of the temperature difference between surface water and deep water more than 1,000 metres (3,300 ft) down. Solar energy heats the surface but that heat seldom gets below about 100 metres (325 ft). So while the surface waters may fluctuate in temperature they are generally always warmer than those deeper down, which stay at a steady 4–5°C.*

It has been estimated that this oceanic temperature gradient could supply many thousands of times the world's current energy needs. Here's how it could work. Surface water is used to boil a volatile liquid; something like ammonia would do. The ammonia vapour then spins a turbine and makes some electricity. Cold water pumped up from the depths is then used to condense the ammonia back into a liquid – and the whole thing starts again.

For the process to work there needs to be a 20°C difference between the surface and deep ocean. Only a narrow band of equatorial waters has this kind of temperature gradient, and to date all attempts to harness this energy have met with failure, the inefficiencies proving too great. However, offshore drilling technologies have been pushed into deeper waters recently, and the lessons learned about

*We are ignoring the polar regions here. The oceanic steam engines would work best in equatorial waters.

working at below 1,000 metres might soon tip the balance back in favour of ocean steam engines. Oil will run out one day, but sea water never will.

The steam engine is old technology for sure, but that does not mean this kind of antique won't feature in future high tech, even in space probes. One day, rovers might patrol the surface of Venus cooled by a refrigerator system invented in 1816.

❄

The Stirling cooler looks simple enough. It's a cylinder fitted with pistons. The pistons move due to the interplay of hot expanding gas and compressed cold gas at either end. There is actually some fiendish complexity in there making it one of the most efficient heat pumps ever devised. The pump can work in either direction – as an engine that transforms heat into motion – or as a cooler that takes the motion of an expanding gas and radiates it away as heat.

The device was invented in 1816 by a Scottish vicar called Robert Stirling. It was largely ignored until the twentieth century, when its great efficiencies were fully realised. Today Stirling coolers are used to keep hot drill heads cool deep underground, and the cryocoolers that keep MRI scanners and other helium-chilled systems cold enough sometimes use the same kind of system.

Basically, a refrigerant is forced into a mesh-like heat exchanger by one of the pistons. As it is compressed, the refrigerant's pressure and temperature rise until a second piston on the far side of the heat exchanger begins to move backwards. The volume and pressure of the warmed gas is now fixed as it moves through the exchanger. This is a network of metal plates with a large combined surface area, and all that metal draws out the heat of the fluid. As a result the refrigerant leaves the heat exchanger at a lower temperature. The first cylinder finds its path blocked by the heat exchangers and so stops moving forwards. The second piston keeps on moving backwards, however, and this draws the fluid into a larger volume making the pressure and temperature drop dramatically. This cold end of the system is used for cooling, drawing in the heat from the surroundings or whatever needs to be kept cold. Finally both pistons move back to their starting positions, all the while keeping the volume between them constant. As they do so, the refrigerant held between them is pumped back through the heat exchanger and warms up again on its way back through.

We are now back at the beginning ready to go again. The heat exchanger, or regenerator, was Stirling's most ingenious achievement. It is a temporary heat store that boosts the cooling effect, making an efficient and easily portable source of extreme refrigeration.

On the surface of Venus, that is what is needed. Venus is the hottest planet in the Solar System. Its soupy atmosphere of carbon and sulphur dioxide shows just how powerful the warming of the greenhouse effect can be. The surface is 450°C – a full 200° hotter than a kitchen oven.

This was all unknown when the first landers made their approaches to Venus in the 1960s. In the popular imagination Mars was seen as an icy place, filled with belligerent beings. Venus was assumed to be some kind of extraterrestrial paradise, blessed with a perpetually warm climate. The space scientists did not entertain these fantasies, of course, but even they were unprepared for the extreme conditions their probes encountered. Spacecraft after spacecraft was crushed in the dense atmosphere. Their plastic components melted and the electronics fried.

The few landers that made it to the surface in one piece had to be built more like pressure cookers than space probes. None of them survived for much more than an hour once safely down. Rovers sent to explore the surface of this volcanic hell would need to have the most efficient air conditioning of any wheeled vehicle. The only working proposition is the Stirling cooler, which would enclose all the electronics inside a ceramic insulator. The hot end would have to reach 500°C to be able to shed heat into the hot Venusian air, while the electronics would be maintained at 200°C, positively chilly by the standards of that planet.

A visit to Venus is not on any of the mission rosters of the world's space agencies. It is undoubtedly a difficult place to work in. But Venus is just as likely to have harboured life in the past as Mars (probably more so). If we decide to find out about that, the Stirling cooler, based on a device invented to pump water from quarries in Scotland, will no doubt play its part.

In 2014, the comet-chasing Rosetta probe dropped a lander, Philae, on Comet 67P/Churyumov–Gerasimenko (shortened to 67P, for understandable reasons). Philae, the first craft to touch down on a comet, was about the size of a small fridge, but if things had been different it would have actually been a fridge. When the mission was conceived back in 1985, the original idea was to land a refrigerator that could keep a sample of the comet's icy mass in pristine condition for an eventual return journey to Earth. It was not to be, and Philae was equipped to analyse the space snowball in situ. Nevertheless, fridges do make it into space for other reasons.

The Hubble Space Telescope and other space-based observation equipment need coolers if they want to look at the heat (not light) coming from the stars. To do that they need to make their heat detectors very cold – less than –200°C – so they pick up the faint heat signatures coming from distant objects.

Crewed spacecraft such as the International Space Station (ISS) need coolants to cope with extremes of temperature as well. In the full glare of the Sun, unfiltered by our atmosphere, the metal components of a spacecraft in Earth orbit reach 260°C. Without a coolant system on board, the ISS would be an unusual place to live. The difference in temperature between the sunny side and the dark side would be 135°C. Finding the habitable spot in between these two extremes would be difficult. So the ISS must shed its excess heat. This is collected by loops of liquid ammonia and other coolants that circulate through the different modules. Those long panels that stick out of the spacecraft are not all solar cells. Some of them are radiators lined with coolant tubes which are designed to let heat dissipate into space.

Future spacecraft might be cooled using a magnetic refrigeration system. This does away with the need for a high-pressure vapour-compression system by making use of something called the magnetocaloric effect. This is an effect in which a suitable material, such as gadolinium metal, becomes warmer as it passes through a magnetic field. That heat radiates away, so the metal is cooler overall once it exits the field, ripe for use in refrigeration.

Magnetic refrigeration systems are becoming available down on the ground. Whether astronauts bound for Mars will get to store their food in a fridge powered by magnets remains to be seen. Refrigeration already has a role to play in pushing back the frontiers of space science. It is being used in the search for dark matter.

❄

Dark matter is mysterious stuff. In the 1930s, Fritz Zwicky and Jan Oort, two astronomers with memorable names, found that galaxies were heavier than they should be. No matter how hard they looked, no one has been able to find this missing 'dark' matter. The search for dark matter has gone into overdrive of late, with several detectors gearing up to be the first to clap eyes on it. The reason why dark matter is 'dark' is because it very rarely interacts with normal matter. Without those interactions we have no way of knowing it is there.

Cosmologists, the scientists who are most interested in this stuff, hypothesise that dark matter might be made up of either WIMPs or MACHOs, or perhaps both. The WIMPs are weakly interactive massive particles. These are weird heavy particles that are proposed to whizz around unseen by our matter-based selves. The MACHO name is a bit more of a stretch: massive compact halo objects. They are dark stars, huge gloomy planets and black holes – big things out in space that are just very hard, if not impossible, to spot with our telescopes.

All the money is on detecting a WIMP and to do that experiments such as LUX (short for Large Underground Xenon) have been set up. LUX is a tank of 368kg (800lb) of liquefied xenon held at -83°C, located 1,500 metres (5,000ft) underground in Homestake Mine, an old gold mine in South Dakota. The mile or so of rock above the tank shields the detector from the noise of cosmic rays that would mess up the experiment if it was done at the surface. The xenon is used because it can be relied on to do nothing at all, nothing that is until a WIMP shows up. Remember, the theory goes that WIMPs hardly ever interact with ordinary matter, like the stuff Earth is made from, so they just sail on through as if the planet was not there. The idea is that sooner or later a WIMP will hit one of the xenon atoms and create a flash of radiation. And the LUX detector will be there to see it. We are still waiting for the flash.

As well as outer space, the application of cold is also taking us to the frontier of inner space. Another staple of the sci-fi cannon is a super-intelligent computer, as clever as any of us, probably more so and even capable of downloading and emulating the entire content of a human brain. We already have an idea of what that kind of computer would be like. It would be a quantum computer, and research into such a device has largely been done at incredibly low temperatures.

*

Readers will be familiar with bits and bytes. These are the units of computer memory. There are eight bits in a byte;

four bits or half a byte is a nibble, while on a larger scale we get into the megabytes and gigabytes with which we are more familiar.

A bit is the basic unit, representing a single piece of information. It has only two states, on or off, open or closed, up or down, go or stop, 1 or 0. The 'digital' in digital computing comes from the digits 1 and 0 used to write the code that controls the computer processor. The processor is a collection of tiny electronic switches, each one storing one bit of information by being switched either on or off.

The dream of quantum computing is to replace bits with quantum bits, or qubits. Instead of being an electronic switch, the qubit gets its on/off state from a quantum characteristic of an atom or other quantum-scale object. That can be something like the spin of an electron which can be in either a high-energy state (spin up) or a low-energy one (spin down). So far, so similar, the bit and qubit sound the same. However, quantum mechanics tells us that an electron spin (or other quantum characteristic) is never certain until it is measured, so every qubit has a particular chance of being either on or off, a 1 or a 0, at the same time.

This is where the fun bit starts. Inside a quantum computer, the atoms that represent the qubits have to be interconnected, or entangled. Two ordinary computer bits, say 1 0, contain two bits of information, but two entangled qubits contain four times as much information: 1 0, 1 1, 0 1, 0 0. Most of the latest desktop computers work by using chunks of 32 bits at a time – a string of 32 1s and 0s. A 32-qubit code contains the equivalent information as 4,294,967,296 1s and 0s. That shows how a quantum computer can be made to process a lot more data, and can pull off unimaginable feats of calculation that leaves even the processing power of supercomputers out in the cold.

So where does refrigeration come in? Well, to entangle qubits effectively they have to be isolated from the hubbub of the rest of the Universe. For a long time, the most promising way of doing that was considered to be cooling the atoms to just above absolute zero so they barely moved at all, which made it possible to manipulate their quantum properties precisely.

Prototype quantum computers are being built inside pulse refrigerators. These coolers work in a similar way to Stirling coolers but generate a pressure wave in the refrigerant, which shoves the heat to the warm end of the system, where it is shed. The pressure wave behaves in a similar way to a sound (or acoustic) wave that our ears can detect. As such it is described as a thermoacoustic wave.*

The pulse refrigerators take the qubits down to just above absolute zero, and it is in these conditions that computer scientists and physicists are figuring out how to entangle large numbers of them to produce the much-vaunted processing power. In 2013, a team working in Canada, the UK and Germany managed to create a coherent quantum computing device that worked for 39 minutes at room temperature. However, they believe that their new technique, which involves using phosphorus ions (atoms that have lost electrons and become electrically charged) as their qubits would be even more stable at temperatures of around 4K.

So what can a quantum computer do? It won't replace the silicon-based one I'm using to type this. If anything, a quantum computer will perform each of its operations more slowly than a classical computer. So what is the big deal

*Heat travels through a superfluid as a wave as well, much like a sound wave which moves with a specific speed. This heat wave exists in no other known material and has been dubbed 'second sound'. The speed of second sound varies with the temperature of the superfluid, which can range from about 4K to close to 0K, and as such measuring that speed is a good way of giving an accurate temperature of the superfluid.

about them? Why bother? The difference will be their ability to juggle immense amounts of data all at once, and that will allow them to perform the so-called 'hard' problems of mathematics. These hard problems are not possible to solve by simply chucking more processing power at them. A classical computer can do its sums very fast, but it does one at a time. To solve a hard problem with a classical computer it would require the computer to perform an infinite number of calculations or operations, or something near to infinite. That obviously takes a long time – something near to the remaining life of the Universe and probably longer. So hard problems are not solvable with computers. That is why they are so hard.

The human brain is able to tackle hard problems because it has some way – currently incomprehensible – of figuring out which operations to ignore and which to focus on. A quantum computer, which is able to do many operations all at the same time, might be able to do what the human brain can do to solve hard problems, and go on to tackle the kinds of calculations that are just too complex for the human mind.

If quantum computing can emulate the human mind, then perhaps something like the human mind, human consciousness, can be made to work inside a quantum computer. The humble refrigerator might one day spawn a new form of machine consciousness. What happens then is anyone's guess.

Such a form of qubit-powered life might be just the ticket for those who seek immortality: as death approaches, simply upload yourself to the Cloud, and avoid having to spend eternity plucking a harp on an altogether different kind of cloud. There are already hundreds of departed folk who are kept on ice, in the hope that they can be downloaded from the afterlife by some future medical technology.

The idea for cryogenic 'life extension' hails from the 1960s, when maths and science teacher Robert Ettinger set up the Cryonics Institute on the outskirts of Detroit (along with a marvellously named charity, The Immortalist Society). History suggests that Ettinger got the idea from science fiction, and history reflects that this is no bad thing. However, he was also inspired by the pioneers of cryomedicine, where bodies are cooled to prevent damage during surgery or to aid with healing.

Today, the bodies – and sometimes just the heads – of more than 100 institute members (eternal membership fee: $30,000) are held in tanks of liquid nitrogen, all preserved in the hours after death to ensure that they are in as pristine a condition as possible. Members all signed up with the hope that the institute would be a temporary graveyard. It is likely to be a very long wait, but they have bought themselves some time. If the freezing process was carried out properly to minimise tissue damage, their lifeless bodies remain as intact as any recently deceased corpse. It is not beyond the realms of possibility that future medics will one day be able to bring back the people stored by the Cryonics Institute and other similar organisations. Nevertheless, if they were reanimated, these people – mostly rather aged when they died the first time around – would surely be at grave risk of dying again more or less straight away.

Until recently, the words 'Frozen' and 'Disney' normally came together during the story of how Walt Disney was cryogenically frozen after death, as opposed to the Nordic saga about estranged princesses. However, it is debatable which story holds more truth. Disney's family are adamant that it is an urban myth, and that their forebear was cremated in 1966. This was well before Ettinger's Institute had opened and a full 10 years before its first inmate – Ettinger's mother – was laid to rest. The second client was Ettinger's first wife, who was frozen in 1987, but from the

1990s business picked up. In 2011, Robert Ettinger himself became Patient 106.

*

By the time medical science has caught up with Ettinger's idea, human society could be beyond our current comprehension. Transport as we know it might have faded from memory as everyone nips about the place by teleportation. It is safe to assume that a quantum computer would be needed for a teleportation machine, the kind of thing that Captain Kirk and Co. use to nip down to the nearest planet. Teleportation of this kind, one that allows large objects to travel without moving, is super science fiction, way down the far end of a futurologist's to-do list. However, in theory what needs to be done for it to work is to capture a snapshot of all the quantum states of all the atoms in the object, transmit that information, then recreate all those states exactly somewhere else.

In 2009, Michio Kaku, the great American science communicator and professional futurologist, characterised the problem of teleportation not as a science problem but as an engineering problem, although 'a very big engineering problem'.

There is indeed a lot to do, but the Bose-Einstein condensate is likely to be involved. Remember, this ultracool substance is so ultracold that the atoms that make it up have lost their individual identity to become a 'super atom', a vibrating wave of quantum material that acts as a single entity. It may one day be possible to cool an entire body so that it transforms into a Bose-Einstein condensate wave, an atomic laser that carries all the information about the original state of the object. This is a big stretch but let's just say that one day someone could figure it out. Certainly 'matter waves' for much simpler objects, such as single atoms, have been produced and teleported across rooms.

For teleportation to be worth doing, the information would need to travel a lot further than across the room. In theory it may be possible to send it to anywhere in the Universe at the speed of light, and on arrival it would be re-formed back into normal matter. To achieve that, one idea being pioneered in Australia is to direct the wave into a separate source of Bose-Einstein condensate (BEC) that is also being illuminated by a laser beam (let's not forget that this is all happening in a super-powerful fridge). As the matter wave enters the BEC, a beam of photons are emitted as the wave merges with the condensate. That beam of photons is effectively a signal that carries information about the original matter wave – information about an atom, a molecule, or all 30 trillion cells of a human body. The signal beam of photons is then sent to another source of BEC at some distant point. As it shines into that condensate, a matter wave identical to the original one is hypothesised to be ejected. That wave can then be converted back into its original 'warm' state (again, someone will just have to figure that bit out).

The original object – including a teleported body – would be destroyed and killed as it plunges into the Bose-Einstein condensate to be converted into a matter wave. At the other end, the matter wave would then create an exact copy (at least that's the idea). A teleported person would have to die first, then be recreated with a new set of atoms.* Philosophers have used this idea already to consider the nature of identity. Engineers would just want to consider how to make everything go back together in the right order.

It sounds far-fetched and that's because it is, but the Bose-Einstein condensate and its close companion the fermionic condensate are opening up these and many other possibilities that have yet to be imagined. Briefly, a fermionic condensate, named after Enrico Fermi, the Italian who was first to

*One small plunge for a man ... One giant plunge for mankind.

harness nuclear fission, is a super-cold substance made from fermions, not bosons. Put extremely simply, bosons do the forces, while fermions do the mass. The first fermionic condensate was produced in 2003.

But what about something tangible, something real-world and possible in our lifetimes? Something like the smart fridge?

*

A smart fridge, or Internet refrigerator, is rapidly becoming an old-fashioned view of the future. It was an early example of the 'Internet of Things' concept. This is where every item in the built environment has some kind of connectivity to the Internet so it can send out useful information into the network and receive commands from human users or artificial intelligence.

The smart fridge was envisioned in the late 1990s, when everything to do with the burgeoning World Wide Web and Internet seemed like a good idea. Basically a smart fridge is able to know what is in it. It might scan the barcodes of the items loaded or make use of radio signals from tiny microchips embedded in the packaging. The fridge knows how long food has been in there and is able to monitor the weight of each item. It then lets its owners know when food has gone out of date or alerts them if they are running low on items. It will learn how often you consume each type of food and use clever algorithms to figure out when it needs to dial up the supermarket and put in an order for favourite items in good time. You can tell the fridge your recipes for the week and leave it to do all the shopping. All you have to do is take things in and out. The fridge does the rest.

Such a device is pretty complicated but perhaps not beyond the realm of possibility. But it is probably beyond the realm of anyone wanting one. In 2000, LG launched the first smart fridge. It cost £13,000, which was a high price tag for a machine that solved problems which could be fixed by just opening the door and having a look at what was inside.

However, as we near the end of our journey through the story of refrigeration, we are in danger of falling into the same kind of trap that snared many of the characters we met along the way. Think of the frigorophobics of 1880s Paris or the Boston merchants who laughed at Frederic Tudor.

The smart fridge might one day just be known as a fridge. One possible future is that the fridge retains its place as the hub of any home, and to do that it will need to get smart in a different way.

It is not uncommon for fridges to have screens on the door that are used as control panels. It is a short step for those controls to evolve into a smartboard, a touchscreen that covers most of the door. We'd use the smartboard to not only control the fridge but also make online orders for food – or send messages, watch the TV or phone friends. The driving force for such technology would not really be about the contents of the fridge, but the content of the entire house.

A smart fridge will be needed for a smart home. If current trends persist, home computers will grow in number but also begin to disappear. They will end up inside everything else in the house. The fridge door is a communication hub already. There are few other empty spaces in a kitchen, and the door may be covered in a collection of post-it reminders, calendars and various lists. A bit of touchscreen tech will put this all online and also turn the fridge into the main control centre for the family. The heating, air con, lighting, laundry, the bath and the shower will all be controlled from the smartboard (and from elsewhere no doubt – there will be an app, for sure). The fridge door will become a window on the world.

*

OK, whether refrigerators create teleporters and artificial intelligence or become the latest must-have computing device is all moot. I'd be happy with any of it. However, there are 1.5 billion people who have no access to refrigeration even today. Perhaps the smartest fridge of all is the Pot-in-Pot system, a solar powered cooler, in the simplest sense, that is within the reach of everyone.

The Pot-in-Pot was designed by a Nigerian teacher called Mohamed Bah Abba in the late 1990s. It is an update of the ancient Arabian *zeer*, an evaporative cooler. Food is packed into a watertight ceramic pot, which is inside a larger unglazed earthenware one. The space between the two is filled with sand, and this lining is soaked in water. The water soaks into the outer pot and evaporates away in the sunshine, drawing heat from the contents of the inner chamber. It's simple but effective. It might just be the next big thing in refrigeration.

Now all is said and done, whether it's a low-tech pot or a high-tech cryogenic cooler, the refrigerator should be hailed as humanity's greatest achievement, jostling for position with the likes of the wheel, printing press and microchip. It earns its place not just for how it has changed the past and made the present, but also for the way it will shape the future.

Further Reading

Among the many sources used to put this book together, those listed below have proved to be the most useful, and they are invariably fascinating and entertaining in good measure.

Buxbaum, Tim (2014). *Icehouses*. Shire Library.

Freidberg, Susanne (2009). *Fresh: A Perishable History*. Harvard University Press.

Rees, Jonathan (2013). *Refrigeration Nation: A History of Ice, Appliances, and Enterprise in America*. Johns Hopkins University Press.

Shachtman, Tom (2000). *Absolute Zero and the Conquest of Cold*. Mariner Books.

Weightman, Gavin (2003). *The Frozen-Water Trade: A True Story*. Hyperion.

Also, I heartily recommend *The Secret Life of Machines*, a TV series presented by Tim Hunkin that was broadcast in the UK between 1988 and 1993. www.secretlifeofmachines.com

Acknowledgements

First and foremost, I would like to thank Sarah for putting up with the many absences in body and in mind that were required to complete this book. Also, thanks to my children for helping to work through the many possible titles. My gratitude to Dr Mike Goldsmith for casting a critical eye over the text and to the subs desk, otherwise known as Mum and Dad. Seasoned writers both, thanks for your hard-won expertise. The book would also not have been possible without the help of Jan and Alan, who provided me with a bolthole from time to time. To Jim Martin and Tim Cooke: thanks for all the fish.

Index